T0230509

# THE FRONTIERS COLLECTION

**Series editors**

Avshalom C. Elitzur
Iyar The Israel Institute for Advanced Research, 76225 Rehovot, Israel
e-mail: avshalom@iyar.org.il

Laura Mersini-Houghton
Department of Physics, University of North Carolina, Chapel Hill,
NC 27599-3255USA
e-mail: mersini@physics.unc.edu

T. Padmanabhan
Inter University Centre for Astronomy and Astrophysics (IUCAA),
411007 Pune, India
e-mail: paddy@iucaa.in

Maximilian Schlosshauer
Department of Physics, University of Portland, Portland, OR 97203, USA
e-mail: schlossh@up.edu

Mark P. Silverman
Department of Physics, Trinity College, Hartford, CT 06106, USA
e-mail: mark.silverman@trincoll.edu

Jack A. Tuszynski
Department of Physics, University of Alberta, Edmonton, AB T6G 1Z2, Canada
e-mail: jtus@phys.ualberta.ca

Rüdiger Vaas
Center for Philosophy and Foundations of Science, University of Giessen,
35394 Giessen, Germany
e-mail: ruediger.vaas@t-online.de

# THE FRONTIERS COLLECTION

*Series Editors*
A.C. Elitzur    L. Mersini-Houghton    T. Padmanabhan    M. Schlosshauer
M.P. Silverman    J.A. Tuszynski    R. Vaas

The books in this collection are devoted to challenging and open problems at the forefront of modern science, including related philosophical debates. In contrast to typical research monographs, however, they strive to present their topics in a manner accessible also to scientifically literate non-specialists wishing to gain insight into the deeper implications and fascinating questions involved. Taken as a whole, the series reflects the need for a fundamental and interdisciplinary approach to modern science. Furthermore, it is intended to encourage active scientists in all areas to ponder over important and perhaps controversial issues beyond their own speciality. Extending from quantum physics and relativity to entropy, consciousness and complex systems—the Frontiers Collection will inspire readers to push back the frontiers of their own knowledge.

More information about this series at http://www.springer.com/series/5342

For a full list of published titles, please see back of book or springer.com/series/5342

Ian T. Durham · Dean Rickles
Editors

# INFORMATION AND INTERACTION

Eddington, Wheeler, and the Limits of Knowledge

 Springer

*Editors*
Ian T. Durham
Physics Department
Saint Anselm College
Manchester, NH
USA

Dean Rickles
Unit for History and Philosophy of Science
University of Sydney
Sydney
Australia

ISSN 1612-3018          ISSN 2197-6619   (electronic)
THE FRONTIERS COLLECTION
ISBN 978-3-319-82904-3          ISBN 978-3-319-43760-6   (eBook)
DOI 10.1007/978-3-319-43760-6

© Springer International Publishing Switzerland 2017
Softcover reprint of the hardcover 1st edition 2016
This work is subject to copyright. All rights are reserved by the Publisher, whether the whole or part of the material is concerned, specifically the rights of translation, reprinting, reuse of illustrations, recitation, broadcasting, reproduction on microfilms or in any other physical way, and transmission or information storage and retrieval, electronic adaptation, computer software, or by similar or dissimilar methodology now known or hereafter developed.
The use of general descriptive names, registered names, trademarks, service marks, etc. in this publication does not imply, even in the absence of a specific statement, that such names are exempt from the relevant protective laws and regulations and therefore free for general use.
The publisher, the authors and the editors are safe to assume that the advice and information in this book are believed to be true and accurate at the date of publication. Neither the publisher nor the authors or the editors give a warranty, express or implied, with respect to the material contained herein or for any errors or omissions that may have been made.

Printed on acid-free paper

This Springer imprint is published by Springer Nature
The registered company is Springer International Publishing AG
The registered company address is: Gewerbestrasse 11, 6330 Cham, Switzerland

*This volume is dedicated to the memory of*
**David Ritz Finkelstein** *(1929–2016).*

# Preface

We have found a strange footprint on the shores of the unknown. We have devised profound theories, one after another, to account for its origins. At last, we have succeeded in reconstructing the creature that made the footprint. And lo! It is our own. (Eddington [1, p. 200])

No element in the description of physics shows itself as closer to primordial than the elementary quantum phenomenon, that is, the elementary device-intermediated act of posing a yes-no physical question and eliciting an answer or, in brief, the elementary act of observer-participancy. Otherwise stated, every physical quantity, every it, derives its ultimate significance from bits, binary yes-or-no indications, a conclusion which we epitomize in the phrase, 'it from bit'. (Wheeler [7, p. 309])

This volume was inspired by an FQXi-funded conference held at Trinity College, Cambridge (March 20–23, 2014).[1] This led to additional work over the subsequent two years that addressed the questions raised at the conference in greater detail. The basic premise of the full research programme was that the 'practical' approach to information (as a resource to be manipulated), while certainly very impressive and fruitful, has radically outpaced our theoretical understanding. Yet, advances in quantum information and other related areas have begun to push the boundaries of the physically possible and suggest new ways of thinking about reality, making a foundational understanding of information increasingly important for future progress and, perhaps, yet more impressive practical outcomes.

Our aim was to take our cue from the later ideas of Arthur Eddington[2] and John Wheeler in an effort to fill in some of the empty theoretical ground. To both Eddington and Wheeler, though their views certainly differed in a great many ways, information occupied a (in fact, *the most*) fundamental position in the order of things. Physical reality, in their view, is shaped by the questions we choose to put to it and is thus built up from the information thus generated. This is the root

---

[1]Details can be found here: https://informationandinteraction.wordpress.com.

[2]By 'happy' coincidence, the conference that inspired this work coincided with the 70th anniversary of Eddington's death.

of Wheeler's famous phrase 'it from bit', which is taken to anthropic extremes in his (rather less famous) 'self-excited circuit' in which all of physical reality (including seemingly paradoxically, ourselves, qua beings of a certain constitution) is determined by the questions of observers whose decisions determine (to some extent) facts as to what has happened and what will happen. The observers (or, less anthropically loaded, the observational equipment involving some irreversible process) play an active, creative role in defining reality[3]:

> 'it from bit' symbolizes the idea that every item of the physical world has at bottom—at a very deep bottom, in most instances—an immaterial source and explanation; that what we call reality arises in the last analysis from the posing of yes-no questions and the registering of equipment-evoked responses; in short, that all things physical are information-theoretic in origin and this is a participatory universe. [9, p. 311]

Observers themselves (and so the *interaction* of observer and observed, or subjective versus objective) also play crucial, though again rather different, roles in their work. Neither of them privileged observers as *external* to physics, but sought to incorporate them in the description provided as ineliminable actors. In Wheeler's case (in his anthropic 'observership scheme'), observers were at the root of existence. In Eddington's case, observers merely 'selected' aspects of reality rather than creating them as such—by including observers we must take seriously the intrusion of certain biases into our mathematical theories of reality such that any theory will contain a fair portion of material that reflects certain of our features as experimenters/observers (under various constraints) rather than objective reality. This leads directly to Eddington's much maligned *a priorism*:

> An intelligence, unacquainted with our universe, but acquainted with the system of thought by which the human mind interprets to itself the content of its sensory experience, should be able to attain all the knowledge of physics that we have attained by experiment. [2, p. 3]

In other words, physics is, for Eddington, partly an exercise in 'observer-spotting': there exist (anthropic) selective aspects that are those parts of a physical theory that are really contributions of the intellect and our constitution as observers and measurers of the world. Hence, there is a kind of anthropic carving of reality in both Wheeler and Eddington that can be distinguished by the powers given to the observers asking the questions and making the decisions: in the former, there is a creation of new facts and in the latter, a selection of facts.

One can gain a better understanding of the connection between Eddington's views and those of Wheeler by looking at the work of Wheeler's student (and long-time friend) Peter Putnam. Putnam was amongst the students who joined Wheeler in Leiden in 1956 when Wheeler was at the beginning of his career in general relativity. Wheeler had supervised Putnam's A.B senior thesis on

---

[3]For Wheeler, observers as passive recipients or 'registers of facts' should be replaced by 'participators': there is no 'ready-made universe' to simply record (see, e.g. [8, p. 286]). Eddingtonian observers are likewise non-passive, though they act on a pre-given objective realm, selecting (not necessarily consciously) certain portions as if those portions exhausted what reality is.

Eddington (completed in 1948 at Princeton), but was not very impressed with his academic abilities [9, p. 254]. While Wheeler spoke of disabusing him of his Eddingtonian belief that the laws of nature can be deduced by pure reasoning, it seems Putnam possessed a certain charisma, and it is not unreasonable to believe that certain Eddingtonian principles trickled into Wheeler's thinking. Putnam apparently did drop Eddington for physics, but later drifted into a bizarre mixture of psychology, linguistics, philosophy, and physics—during this time, as Wheeler relates, Putnam was making money as a janitor (rejecting his family's considerable wealth). But Wheeler remained friends with Putnam until Putnam's death in the late 1970s.

In his large, unpublished manuscript 'Comments on Eddington' (from 1962, with a preface added in 1971[4]), Putnam integrates the Wheeler–Everett many worlds interpretation with Eddington's *Fundamental Theory* and his general philosophy, as outlined in *The Philosophy of Physical Science*. In the preface, Putnam writes that he spent a decade (1962–1972) figuring out how to embody an Eddingtonian lifestyle (a 'world outlook'). It seems that Putnam was concerned with getting a physics that could cope with the 'felt real' (with the experience of observers). He wanted a universal science—indeed, perhaps some of the problems people have had with Eddington's later work is that they are trying to force it into the mould of physics, when some of it clearly is not quite physics. Putnam is certainly correct in saying that Eddington's approach is 'not physics in its usual sense' (p. 2), and he in fact viewed its significance as lying in the treatment of the brain (or observer) as a parallel calculating machine: that is, as **a device that deals in information**. In other words, it is sub-physics, and is therefore supposed to say something about the operations that physicists themselves carry out: physics (*the* fundamental theory in Eddington's sense) involves, then, a study of the operations of the calculating machine (us) and has information and interaction at its centre. Hence, Eddington was not concerned with a description of the world independent from observers (the 'absolute world'), but with the description of our means of interacting with it and gaining information: the observer could not be removed from scientific practice.[5] Putnam believed that this incorporation of the observer into physics resolved certain philosophical problems that led to considerable conceptual confusion. This centrality of the observer also brings information and interaction centre stage, as we have discussed, and that is this book's focus.

---

[4]Downloadable at: http://peterputnam.org/comments_on_eddington_1962_preface_1971.htm. It seems Putnam could not let Eddington go so easily. However, the essay is by no means hero-worship, and Eddington takes much flack in this work.

[5]Whether this was in fact Eddington's view is not what concerns us here. Rather, it is the link to Wheeler's 'it from bit' idea. The link seems to be especially strong when it comes to the measurement process, and the idea that this should itself be part of physics so that the subjective and objective are reconciled. Putnam, as mentioned above, related this directly to the Everett–Wheeler view in which the measurement process itself (involving human observers, as a matter of fact, but not in any supernatural way) should be included as part of physics.

Wheeler pushed the notion of measurement and the role of the observer into territory easily as uncharted as Eddington. In the context of the delayed-choice experiment, for example Wheeler envisages a scaled-up version involving photons having travelled a billion light years from a quasar, separated by a 'grating' of two galaxies (to act as lenses offering two possible paths for the light), to be detected at the Earth using a half-silvered mirror at which the twin beams can be made to interfere. For Wheeler, this means that the act of measurement (our free choice) determines the history of that entire system: actions by us NOW determine past history THEN (even billions of years ago, back to the earliest detectable phenomena, so long as we can have them exhibit quantum interference). It is from this kind of generalization of the delayed-choice experiment that his notion of the Universe as a 'self-excited circuit' comes: the Universe's very existence as a concrete process with well-defined properties is determined by measurement. Measurement here is understood as the elicitation of answers to 'Yes/No' questions (e.g. did the photons travel along path A or B?): bit-generation (gathering answers to the yes/no questions) determines it-generation (the universe and everything in it). However, Wheeler's notion does not privilege human observers, but rather simply refers to an irreversible process taking uncertainty to certainty.

The Eddington–Wheeler link can be made a little more precise through the notion of *idempotency*. An operator A is said to be idempotent if $A^2 = A$. Idempotent operators have eigenvalues of 1 and 0, corresponding algebraically to yes–no (bit-based) logic. Eddington employed idempotency at a somewhat deeper level (superficially, at least) than Wheeler, attempting to latch on to the fundamental structure of physics (and knowledge). Eddington used idempotent operators to mathematically define a notion of *existence* (and non-existence). In this way, Eddington defined the elemental structure of reality (true, basic individuality), and it maps closely onto Wheeler's understanding: what is elemental is the yes/no logic (to be or not to be, that, in this case, is the answer!).

The phrase 'it from bit' can be a little misleading, then, since it suggests something static and eternal: whatever *is* (i.e. that which exists) is made from information. But that does not capture what is really going on. The idea embodies a *creative* principle. The settling of questions about the quantum world via measurement interactions creates facts about the world. There is here a curious amalgamation of Bohr's teachings with Eddington's. We think that these links, and the deeper meaning of 'it from bit' (as the genesis of reality, or facts, from observership), have yet to be explored sufficiently and hide many more secrets.[6]

It is worth noting here that, though Eddington's later work was much maligned in its day, and continues to be misunderstood by a great many authors, many of his ideas have proven to be particularly prescient in retrospect. For instance, the 'statistical' portion of *Fundamental Theory*, comprising the first six chapters, is essentially one of the earliest attempts to develop a rigorous theory of reference

---

[6]As Helge Kragh explains in his chapter, Eddington restricts the 'its' to the laws and constants (pure numbers) of physics.

frames that incorporates elements of quantum theory with the geometric spirit of relativity by applying quantum mechanical uncertainty to the origin of spatiotemporal reference frames. There has recently been a resurgence of interest in the relationship between quantum mechanical principles (including the uncertainty relations) and the geometrical aspects of reference frames, most notably within the fairly new field of relativistic quantum information.

More specifically, Marco Toller began considering some of these very ideas in the late 1970s [6], many of which were later taken up by Carlo Rovelli who incorporated them into his relational interpretation of quantum mechanics [4, 5] which is decidedly Eddingtonian in spirit. Peres and Scudo even considered how one might utilize a quantum system to transmit information about a Cartesian frame [3], which is spiritually very close to Eddington's statistical ideas. This raises the issue of one particularly misunderstood aspect of Eddington's work. Superficially, if one only engages with his *Fundamental Theory* or *Relativity Theory of Protons and Electrons*, one might be tempted to assume that Eddington attached an certain objective/ontological status to some of his conclusions. If one, however, studies his *Philosophy of Physical Science*, it becomes evident that his view is decidedly subjective and epistemological. Indeed, he often refers to his view as 'selective subjectivism'.

While Eddington was relatively isolated towards the end of his career, Wheeler continued actively collaborating throughout his life. In fact, he actually worked directly with one contributor to this volume (Wootters). As such, his views on these topics have benefited from contemporary analysis and clarification, i.e. from an active engagement by the physics community while he was still formulating his views. Eddington, by contrast, had very little contemporary engagement with his theories. In some ways, this was because Eddington was simply well ahead of his time.

In any case, the chapters that follow engage with many of these same deep issues of information and/or observership interaction that float atop them: often we find that (what were thought to be) objective/ontological features of the world are bound together with subjective/epistemological features of observers.

That being said, Durham, in Chap.1 of this volume, argues that for science to have succeeded so dramatically, some objective aspects of reality must exist (in a later chapter, Fuchs argues that QBism actually allows for some level of realism). He argues that this objective aspect is embodied in physical laws themselves which are inherently relational in nature. This may seem counter-intuitive given that we have just argued that relational structures are subjective/epistemological, but Durham makes the case that some of those relational features are actually objective. Mathematics, being the way in which we describe these laws, is thus best viewed as a language of description rather than in a purely Platonic or formalist light. Durham further argues that mathematics actually arises from physics rather than the other way around, which bears some similarity to constructor theory, as described in Marletto's chapter. In this sense, Durham places Wheeler's 'it from bit' in the context of mathematical representation while holding to Eddington's dictum that we not lose sight of the original physical or logical insight that led us to a particular mathematical deduction in the first place.

Fully understanding the nuances in Eddington's thinking, however, requires going a bit deeper than mere science. Indeed, as Putnam has noted, Eddington's work was more than mere physics. Stanley, then, examines some of the modes of thought, both about science and religion, that prove useful in understanding Eddington's thinking in this regard. In particular, he emphasizes the importance of experience to both science and religion for Eddington who was a devout Quaker. Quakerism, unlike many other religions, is notably rooted in the non-dogmatic experience of the individual and thus, like physics, places the observer at the centre.

Of course, there is the traditional view that sees Eddington's *Fundamental Theory* (which, it should be noted, was posthumously assembled from his collected notes) as an early attempt at a 'theory of everything' (unifying quantum physics and relativity, as well as the large and small). Kragh offers a wide-ranging discussion of Eddington's work within this context. As he explains, Eddington's fundamental theory was so constructed that the laws and constants, derived as they were from epistemological considerations, could not be ruled out experimentally. But that was precisely the problem that most physicists had with it!

Like Kragh, Rickles' essay deals with the interpretation of Eddington's subjectivism and *a priorism*, taking his lead from Edmund Whittaker's[7] more lenient approach, focusing on the underlying general principles, rather than the specific applications in the computations of pure numbers of physics. He attempts to show how some sense can be made of them if cast in more modern terms, issuing primarily from quantum gravity research and the discussion of observables. Rickles shows that the more radical epistemological claims are not quite right, since observation (or a *plan* of observation/measurement by an observer) still lies at the root of Eddington's version of a priori knowledge: a priori is, for Eddington, tantamount to 'observer-spotting'.

But this volume is more than merely a reanalysis of Eddington and Wheeler themselves. It is an exploration of the ideas and spirit that their work embodied. As such, Weinert proposes that both local and cosmic arrows of time are theoretical constructions (inferences) from available information (criteria), in complete agreement with the Eddington–Wheeler epistemological approach. While attempts to identify the arrow of time (or arrows of time) with particular physical processes have often led us astray, Weinert argues that there are numerous criteria which allow us to infer the anisotropy of time.

Chiara Marletto then lays out the constructor theory (jointly developed with David Deutsch). There are many elements that bear close resemblance to features of Eddington's philosophy in this approach, not least of which is the use of 'principles'. In the case of Eddington, we find that the qualitative laws (relating to our constitution as observers) have the form of prohibitions on certain operations (what

---

[7]It was Whittaker who assembled Eddington's notes after his death and published them as *Fundamental Theory*.

Edmund Whittaker refers to as 'postulates of impotence'—see Rickles' chapter). In developing constructor theory, Marletto addresses a curious aspect of information (as ontology): it seems to be rather an abstract thing, independent (to a large extent) from specific physical objects (what she calls 'substrate independence of information'). This is in contrast to Wheeler's view that the information, in the form of bits, literally *was* the substrate of universe. The fundamental principle of constructor theory, then, is that every physical theory is expressible via statements about what physical transformations (tasks) are possible (or not possible) and why. In other words, counterfactual statements are taken to be fundamental as opposed to factual statements.

By way of a selection of relevant correspondence, Chris Fuchs tackles aspects of Wheelerian observership and takes seriously the idea of 'No participator, no world!' As Fuchs points out, Qbism (the approach Fuchs has pioneered with a variety of collaborators) owes much to Wheeler's later thoughts on the meaning of quantum mechanics. Fuchs battles with similar foes to Eddington and Wheeler before him: if agents/observers are central to the interpretation of the theory, then doesn't it lose it's realist character? Isn't it too subjective? Isn't it instrumentalist, positivist, or perhaps operationalist? Fuchs argues that the intrusion of 'the subjective factor' (into the interpretation of quantum states) by no means pushes the view into anti-realism: including the observer ought not to be tantamount to anti-realism. Indeed, this is akin to Eddington's 'selective subjectivism': objective reality exists but our knowledge of it is subjective (contrast this with Durham's views as outlined in Chap. 1).

Constructor theory and QBism can then be contrasted with a more operational 'agent-based' (or 'task-based') approach to physics ('resource theory') presented by Younger-Halpern, in which the experimentalist and their freedom to act on and influence a physical system (under some constraints), are centre stage. For example, a classical agent would be constrained to perform only local operations and would thus have a hard time creating entangled systems. The restrictions on agents' abilities (what kinds of operations they can carry out) determine theories from this resource perspective. Younger-Halpern lays out the basic project and advances various challenges and possibilities for the approach, including (as did Fuchs, in a different way), the issue of scientific realism with respect to such approaches.

As we mentioned earlier, this volume benefits from the fact that one contributor—Wootters—worked directly with Wheeler. In the past several decades, it is then no surprise that Wootters has been one of the leaders in the struggle to 'reconstruct' quantum mechanics from first principles. The approach he takes in this particular work is based on the fact that in order to do physics, we need to connect the entities that appear in our physical theories to the objects that we experience in the actual world. This ties in nicely with some of the earlier chapters that emphasize the experiential nature of reality. The core of Wootters' argument is a toy model that gains a non-trivial probabilistic structure from the imposition of a principle that he called 'the maximization of predictability'. In a way, he echoes Leibniz's assertion that we live in the best of all possible worlds, i.e. the world is best in the sense of being the most predictable in the face of an underlying randomness.

Though he never worked directly with Wheeler, Knuth was partly inspired by him to reconsider what one might usually think of as a paradigm of elementarity and objective individuality: the electron. He develops a view ('Influence Theory') according to which the relationship of the observer and the observed is far more widespread than, for example, notions of length measurements in the context of special relativity. Many of the (supposedly intrinsic) properties of the electron are argued to be relational, holding between the observer and the observed system (the electron), with a simple direct, discrete influence relation mediating their interaction. Hence, the respective shares played, in our physical description of the world, by observer and observed is, according to this model, not quite as simple as most simple accounts of scientific representation would have us believe (as with Eddington's, Wheeler's, and others from this volume).

This collection of papers only scratches at the surface of the rich body of work that exists and that shares a singular spiritual (and sometimes direct) kinship to the work of two titans of twentieth-century physics. Indeed, Eddington and Wheeler provide appropriate bookends for the greatest century in the history of physics: in many ways, they held similar views while in others they were diametrically opposed. On the one hand, both insisted that the role of the observer was crucial to our understanding of physics. On the other hand, it could be argued that Wheeler did not believe in an objective reality while it was quite clear that Eddington did.

Either way, it is our hope that this volume will stimulate further discussion and debate concerning the role of the observer in physics, the nature of our subjective relationship with the universe, and the nature and role of information in regards to both. As Wheeler once said,

> At the heart of everything is a question, not an answer. When we peer down into the deepest recesses of matter or at the farthest edge of the universe, we see, finally, our own puzzled faces looking back at us. (John Wheeler, as cited in John Horgan, The End of Science, p. 84 (Little, Brown & Company, 1998).)

Manchester, USA                                                                  Ian T. Durham
Sydney, Australia                                                                 Dean Rickles
May 2016

# References

1. Eddington, A.S.: Space, Time, and Gravitation. Cambridge University Press (1920)
2. Eddington, A.S.: Relativity of Protons and Electrons. Cambridge University Press (1936)
3. Peres, A., Scudo P.F.: Transmission of a Cartesian frame by a quantum system. Phys. Rev. Lett. **87**, 167901 (2001)
4. Rovelli, C.: What is observable in classical and quantum gravity? Class. Quantum Gravity **8**, 297 (1991a)
5. Rovelli, C.: Quantum reference frames. Class. Quantum Gravity **8**, 317 (1991b)
6. Toller, M.: An operational analysis of the space-time structure. *Il Nuovo Cimento B* **40**, 27–50 (1977)

7. Wheeler, J.A.: Information, physics, quantum: the search for links. In: Zureck, W.H. (ed.) Complexity, Entropy, and the Physics of Information. Addison Wesley, Redwood City (1990)
8. Wheeler, J.A.: At Home in the Universe. By John Archibald Wheeler. American Institute of Physics Press, Woodbury, New York (1994)
9. Wheeler, J.A.: Geons, Black Holes, and Quantum Foam. W. H. Norton (2000)

# Acknowledgements

Thanks to the FQXi (Foundational Questions Institute) for providing the grant that enabled the original workshop that inspired this volume to take place. Thanks also to St. Catharine's and Trinity Colleges, Cambridge, for providing a space for the open discussion of these ideas. ITD would like to thank Laura Bellavia for administrative support and Saint Anselm College for additional financial support. DR would like to thank the Australian Research Council for financial support via grant FT130100466 and the Future Fellowship on which this book was completed and also to the John Templeton Foundation for financial support for the project 'New Agendas for the Study of Time', which contributed to the current volume. Thanks also to Max Kemeny for his excellent editorial work.

# Contents

# Contributors

**Ian T. Durham**  Saint Anselm College, Manchester, NH, USA

**Christopher A. Fuchs**  Department of Physics, University of Massachusetts Boston, Boston, MA, USA; Max Planck Institute for Quantum Optics, Garching, Germany

**Kevin H. Knuth**  University at Albany, Albany, NY, USA

**Helge Kragh**  University of Copenhagen, Copenhagen, Denmark

**Chiara Marletto**  Materials Department, University of Oxford, Oxford, UK

**Dean Rickles**  University of Sydney, Sydney, NSW, Australia

**Matthew Stanley**  New York University, New York, USA

**Friedel Weinert**  Bradford University, Bradford, Yorkshire, UK

**William K. Wootters**  Williams College, Williamstown, MA, USA

**Nicole Yunger Halpern**  Institute for Quantum Information and Matter, California Institute of Technology, Pasadena, CA, USA

# Chapter 1
# Boundaries of Scientific Thought

Ian T. Durham

## 1.1 Introduction

The scientific revolution, as understood to be the rise of modern science, began in the late Renaissance and took firm hold during the Enlightenment. Indeed, it played an integral role in general Enlightenment-era thinking with its emphasis on empiricism and rational thought. It is no accident that this period in history was dominated by classical 'Renaissance men' (who, despite the term, were hardly confined to the Renaissance—a prototypical example is Benjamin Franklin). The modern concept of 'disciplines' in which a person becomes an expert in only one, narrowly defined area whose methods tend to be confined to that field, did not yet exist. Science was ultimately a methodology that was meant to apply to anything. Indeed, many Enlightenment-era thinkers did not presume that there were any limitations to its applicability. It was merely a reliable way to approach the world in general, incorporating rational thinking, empiricism, and mathematics.

One of the key tenets of this early period of scientific thought was the assumption that there exists an objective reality and that it is the task of science to uncover that reality. The 20th Century brought about radical new discoveries that challenged the assumption of an objective reality. This had been preceded by the rise of individual academic disciplines at universities in the late 19th Century. In the process, scientific methodology became associated, almost exclusively, with what are now recognized as 'the sciences.' Further segmentation has led to 'sub-methodologies' that tend to only be applied in narrow sub-disciplines. The combination of the fragmentation into strict disciplines with the challenges presented to objective reality have caused us, to some extent, to lose sight of the underlying framework that has made modern science arguably humanity's greatest achievement. Unfortunately, this also comes at a time when science and rational thought itself are increasingly under assault by a

I.T. Durham (✉)
Saint Anselm College, Manchester, NH 03102, USA
e-mail: idurham@anselm.edu

© Springer International Publishing Switzerland 2017
I.T. Durham and D. Rickles (eds.), *Information and Interaction*,
The Frontiers Collection, DOI 10.1007/978-3-319-43760-6_1

dramatic rise in motivated reasoning and a general desire for simplistic solutions[1] to complex problems.

In this essay I address some of these problems, with a particular emphasis on physics which is the core scientific discipline. Indeed, while not everything in the universe must obey the laws of biology (e.g. a rock) or even the laws of chemistry (e.g. a weak nuclear interaction), absolutely everything must obey the laws of physics. And so, through the lens of physics, I define the boundaries of scientific thought. In Sect. 1.2, I give a broad argument in favor of the existence of a true objective reality and, in the process, attempt to correct several pernicious misunderstandings about certain aspects of physics. In Sect. 1.3, I eschew the usual formalist versus Platonist debate in favor of a third alternative, arguing that mathematics is a *formal language* by which we describe the objective reality that I argued for in Sect. 1.2. Putting these ideas together, I argue in Sect. 1.4 that the structure of objective reality is in the physical laws themselves which are inherently relational in nature. I further argue that we often mis-represent how these physical laws develop, what they mean, and what they have to say about other, related physical laws. Finally, in Sect. 1.5 I note that many of these 'laws' actually place inherent limits on what we *can* know about the world and that we often over-extrapolate and over-interpolate our results.

## 1.2  Objective Reality

The first few decades of the 20th Century witnessed the introduction of two of the greatest scientific developments in history: relativity theory and quantum theory. Despite the fact that the two theories appear to be at odds over many things, both seem to suggest that reality is subjective. This, of course, has had a notable influence on our larger culture and notably on the post-modernist movement which, despite the derision with which it is often rightfully treated, has had a profound impact on society at large. While the question I wish to address here is one of pure science and not one of philosophy or culture, I hope that my arguments will have an impact beyond pure science.

### *1.2.1  Relativity*

Let us consider relativity theory first. To motivate our investigation, I pose the following question: what is the most important lesson we can take from relativity?

---

[1] I prefer the term 'simplistic solutions' to 'simple solutions' since, to me, the former invokes solutions that are easy or less disruptive to some pre-conceived set of ideas we might have. In other words, some complex problems do indeed have simple solutions, but they may not be the solutions we desire because they challenge our worldview. Thus what people often really seek are simplistic solutions rather than simple ones.

It is my entirely unscientific guess that most people—scientists and non-scientists alike—would say that relativity's most important lesson is that the values of certain physical quantities depend upon the reference frame in which they are measured. This strongly suggests that relativity's most important lesson is that reality is subjective. Instead, I contend that relativity's most important lesson is *the exact opposite*. To understand this, let us first consider a bit of history.

The idea that motion is relative is very old. The Ancient Greeks certainly had a basic understanding of the phenomenon as did the Chinese at least as far back as the Jin dynasty[2] in the third and fourth centuries CE [13]. It was Galileo who first proposed that this had an implication for physical laws. Indeed, it was Galileo in his *Dialogo sopra i due massimi sistemi del mondo* (*Dialogue Concerning the Two Chief World Systems, 1632*) who first introduced the principle upon which all formal relativistic theories are founded. Formally it is stated as follows.

**Definition 1.1** (*Principle of Relativity*) The laws of physics must be the same in all inertial reference frames.

This principle was one of the core assumptions made by Einstein in his development of special relativity. To that he added the additional assumption that the concept of 'speed' had a finite, upper limit whose value was equal to the speed of light. As such, the speed of light took on the status of a 'law of physics' (since it served as the upper bound on all speeds) which meant that it was, by the first assumption, necessarily the same in all inertial frames. All of special relativity can be deduced from these two basic assumptions.

Galileo's deduction of this principle was based on a thought experiment involving a sailor in a windowless room (belowdecks) on a ship sailing in a straight line at a constant speed on perfectly calm water. He reasoned (correctly) that there was no experiment the sailor could perform without leaving the room that could prove whether or not the ship was moving. Hence motion must be a relative (subjective) concept. None of Galileo's contemporaries took this to mean that reality itself was subjective. Quite the opposite, they correctly understood that it meant that the actual *laws* of physics must be the same whether the ship was moving or not. Thus while certain physical quantities may indeed be subjectively dependent on a reference frame, the *laws* governing how those quantities are related and how they are obtained, must be entirely *in*dependent of any reference frame. But if they are independent of any reference frame then they must be objective.

Though Einstein made Galileo's principle a core axiom in his development of special and general relativity, the conclusions he drew were far enough removed from personal experience that they spawned numerous misconceptions in the popular imagination. As an example, consider the famous Twin Paradox of special relativity. Suppose that we have two twins, Alice and Bob, and let us assume that Alice becomes an astronaut who travels at a highly relativistic speed to a distant star before returning. Relativity tells us that time runs slower for observers that are in motion. Thus, Bob expects that when Alice returns she will be younger. But relativity also tells us that

---

[2]This is not to be confused with the "Great Jin" which was a dynasty in the twelfth century CE.

motion is relative and so, from Alice's standpoint, it is Bob (and the entire solar system, for that matter) that moved while she remained stationary. Thus she expects that *he* will actually be younger. The paradox is, of course, that they can't both be right. For reasons lost to history or that are a matter for psychologists and sociologists, the paradox, as opposed to its solution, has always been the focus of the narrative. While most physicists know that there really is no paradox here, this notion that "everything is relative"—which is a byproduct of the paradox and not it's solution!—has seeped back into physics from the popular culture, missing the fact that popular culture misappropriated the idea to begin with. We forget the true moral of the tale: when Alice actually does return to greet her brother, *there can be only one reality*. We can't even conceive of a reality in which both of them are right.[3] So while they may find different values for certain observables in their respective frames, the *laws* of physics must be immutable, i.e. frame-independent. Operationally this implies that there must be a procedure for reconciling their measurements.

This should make sense if we pause to think about it. What is the use of a physical theory if it doesn't tell us something widely applicable about the world? To put it another way, we rely on the fact that the world works the same way at the physical level in New York as it does in Tokyo. It is impossible to even imagine a world in which it didn't. It is *this* point that is the essence of relativity. Far from telling us that everything is relative, relativity actually tells us that there are certain things (physical laws) that *must* be universally true and absolute. In short, far from suggesting that "everything is relative" (i.e. subjective), relativity explicitly *requires* some level of objective reality. This, in fact, serves as a core idea in Einstein's general philosophy of physics: **the universe is only comprehensible if there is something about it that can be objectively known**. Indeed, we instinctively rely on this objectivity every day in our expectation (consciously or unconsciously) that physical laws are universal, i.e. that our confidence in engineering and even such things as simple as getting out of bed in the morning, is not misplaced.

One final note concerning relativity is that many physicists who do understand the theory may be tempted here to assume that objectivity necessarily implies determinism. Indeed, I suspect it may be a key component in the thinking of ardent defenders of local realism. I will not claim that is the case here (nor will I deny it), though the relationship between determinism and objective reality is worthy of further examination.

## 1.2.2  Locality

The question of an objective reality becomes a bit murkier in light of certain aspects of quantum theory. We make two major assumptions in classical physics that prove problematic at the quantum level. The first that I wish to discuss is locality.

---

[3]The solution to the paradox, of course, is to analyze the problem from a third reference frame (often chosen to be centered on the sun). In doing so we find that it is Alice that is actually younger when they meet again.

**Definition 1.2** (*Principle of Locality*) Physical systems are only directly influenced by their immediate surroundings.

This statement of the principle is a bit misleading since it implies a certain spatial "closeness." Historically the principle arose as a counter to "action at a distance" by suggesting that for an action at one point in spacetime to have an influence on some other point in spacetime, something must exist (such as a field) in the spacetime between the two points that mediates that influence. In other words, for object $A$ to exert an influence on object $B$, something *physical* must travel through spacetime between $A$ and $B$ in order to carry the influence from one to the other. It bears noting that this is the definition of an interaction in the Standard Model which predicts four fundamental interactions (electromagnetism, gravity, weak nuclear, strong nuclear) plus the Higgs interaction.

Locality is usually assumed to hold unequivocally in classical systems and is an axiom of relativistic quantum field theory. But it can prove problematic when we try to make sense of certain types of physical systems. For example, all matter is stable against collapse under the enormous electrostatic forces that bind atoms and molecules together. This stability is a result of degeneracy pressure[4] which is not traceable to any fundamental interaction but rather to the spin-statistics connection. As A. Zee puts it

> It is sometimes said that because of electromagnetism you do not sink through the floor and because of gravity you do not float to the ceiling, and you would be sinking or floating in total darkness were it not for the weak interaction, which regulates stellar burning. Without the spin statistics connection, electrons would not obey Pauli exclusion. Matter would just collapse [63].

Nothing (such as a field) mediates the spin-statistics connection. It simply is. Therefore degeneracy pressure must be a non-local phenomenon.

Non-locality is not in-and-of-itself problematic for objective reality, per sé, but it does pose some problems for causality and determinism. It is worth noting here that causality and determinism are not necessarily the same thing. D'Ariano, Manessi, and Perinotti, for example, have developed a toy operational theory that is deterministic but not causal. In order to understand the implications this has for objective reality, it is important to take a closer look at the difference. Consider any probabilistic physical theory. As defined by D'Ariano et al. such a theory is *deterministic* if all of the probabilities of any physical events are either zero or one [12]. If the probabilities are anything other than one or zero then we might say the theory is *partially* deterministic.

First consider the following simple situation. A blue index card and a red index card are each placed in separate envelopes and the envelopes are sealed. One of the envelopes is then opened and found to contain the blue index card. That means, of course, that the other envelope must necessarily contain the red index card regardless of when (or even if) it is opened. Thus, in a certain sense of the word, the act of opening the first envelope and inspecting the color of the card inside *determines*

---

[4]For a concrete example, see [16].

what we would find if we opened the second envelope. The same could be said had we opened the envelope containing the red index card first. Of course, this assumes that the blue card is always blue and the red card is always red, i.e. it takes realism for granted (something I discuss further below). In other words, there's nothing particularly odd about this example. Nevertheless, knowledge of one card's color immediately determines knowledge of the other even though the acts of opening the envelopes could be spacelike separated events. That, of course, is the key here: determinism is a description of states of *knowledge* about a physical system, not a description of the actual physical system itself. As such, locality is not a requirement for a deterministic system. Indeed, degeneracy pressure is deterministic but not local. But can a physical system be *causal* and non-local? Before we address that question, we need to discuss realism.

### 1.2.3 Realism

The second assumption that is made in classical physics that proves problematic at the quantum level is *realism* which is the idea that physical quantities are real, i.e. they exist independent of whether or not we measure them. So, for example, in Schrödinger's famous feline thought experiment, classical physics assumes that at any given instant prior to opening the box, the cat is either alive or dead, i.e. it has a definite state that is independent of the box's state. To put it another way, consider the old philosophical thought experiment[5] "if a tree falls in a forest and no one is there to hear it fall, does it make a sound?" The realist answer to this question is an unequivocal 'yes.'

Realism proves to be problematic in some quantum systems, partly due to the fact that quantum systems are *contextual*. Contextuality is usually understood as placing limitations on the results of certain quantum measurements based on observables that commute with the observable being measured. It can also loosely be interpreted as saying that quantum measurements typically depend on the context within which they are measured [18]. So, for example, consider a sequence of spin-$\frac{1}{2}$ measurements on a certain quantum system, as shown in Fig. 1.1. Realism assumes that the spin has a definite value along a given axis whether or not it is measured along that axis, e.g. if the spin is measured along, say, axis $A$ and was found to be aligned with that axis, then regardless of any subsequent measurements along other axes, a second measurement along $A$ must necessarily show the spin to be aligned with that axis. In other words, realism would appear to suggest that once it is aligned with an axis, it is always aligned with that axis regardless of any subsequent measurements made along other axes.

---

[5]The origins of this question appear to be in George Berkeley's *A Treatise Concerning the Principles of Human Knowledge* (1710). Its current form seems to have first been stated in [39].

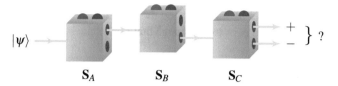

$|\psi\rangle$ ———

$S_A$  $S_B$  $S_C$

$+$ $-$ $\}$ ?

**Fig. 1.1** Each *box* represents a measurement of the spin for a spin-$\frac{1}{2}$ particle along some axis with the *top* output indicating that the state is aligned ($+$) with the measurement axis and the *bottom* output indicating that the state is anti-aligned ($-$) with the measurement axis. *Red* and *blue* lights on the *top* simply indicate to the experimenter which of the two results is obtained (e.g., *red* might indicate aligned and *blue* might indicate anti-aligned)

On the other hand, quantum theory tells us that the probabilities associated with the two possible outcomes of a measurement along axis $C$ in the Fig. 1.1 solely depend on the relative alignment of axes $B$ and $C$ and the state entering the device that measures along $C$. For example, in the figure, the state exiting the middle device measuring axis $B$ is $|b-\rangle$ where b represents a measurement basis. The probabilities for the outcomes from the third device are thus $\Pr(c+) = \sin^2 \frac{1}{2}\theta_{BC}$ and $\Pr(c-) = \cos^2 \frac{1}{2}\theta_{BC}$, where $\theta_{BC}$ is the angle between the $B$ and $C$ axes [48]. Now suppose that $\theta_{BC} = \frac{\pi}{2}$. This means that $\Pr(c+) = \Pr(c-) = 0.5$ meaning the, state as measured along axis $C$, could equally well be aligned or anti-aligned with that axis. *This result is entirely independent of the outcome of any previous measurement.* But that, then, means that if $A$ and $C$ represent the same axis, it is possible that the state will initially be $|a+\rangle$, but then later be found to be $|c-\rangle = |a-\rangle$.

Realism would require that if the particle is found to be aligned with the $A$ axis then it would *always* be aligned with that axis unless some action is taken to change that. Classically, if we measure the spin angular momentum around one axis, there should be no change to the spin angular momentum around any other axis. Indeed aerospace engineers generally take this as axiomatic when working with pitch, roll, and yaw. So no measurement subsequent to the measurement along $A$ should change the spin state of our particle in reference to $A$ and yet it may. This would suggest that there is a problem with realism on the quantum level and implies that the reality of certain quantum measurements is *subjective*, i.e. it depends on the context in which the measurement is made.

This is the second misconception about the subjectivity of physical laws and procedures. Notice that there is a qualification here. Only *certain* quantum measurements behave in this manner. In the example given, while the spin state is contextual, the fact that the particle is a fermion is *never* in question. In fact it is not even conceivable that it could be otherwise since the device in question is only designed to measure fermions.[6] Or, to put it more bluntly, a measurement of the spin state of an electron will not produce a potato. Ultimately we could attribute this behavior to a

---

[6]One could design an experiment that measures the spin of both fermions and bosons, but that would be a different experiment than the one described.

kind of conservation law of sorts regarding the nature of the physical system under investigation. Indeed, it is persuasive enough that I will state it as a general principle.

**Definition 1.3** (*Principle of Comprehensibility*) The nature of a physical system under investigation will always remain within the bounds of the method of investigation.

I use the term 'comprehensibility' here because the point is that we expect that we will, in some way, comprehend the results of our observations and measurements. Even if we don't fully understand them immediately, we expect that, eventually, science will offer an explanation. To put it another way, we expect scientific answers to scientific questions. So if Wheeler is correct and the universe is simply built up from answers to yes/no questions, then the principle maintains that no question we put to the universe will ever produce anything but 'yes' or 'no' (or some combination of the two). For example, consider an idealized Elitzur-Vaidman bomb test [22]. In that thought experiment, an interferometer is used to test whether or not certain bombs are duds. In such an interferometer, there are two photon detectors, call them D0 and D1. In the idealized model, these photon detectors have perfect efficiency, i.e. every photon that enters the interferometer is assumed to register at one of the two detectors. A bomb is placed in one beam of the interferometer such that if an entering photon is detected by D1, then the bomb is known to work. If the photon reaches detector D0, it is inconclusive and further tests must be performed. But it is never in question that the photon reached D0 in the idealized model. The inconclusiveness is related to how we *interpret* the result, not in whether there actually *is* a result to be interpreted.

So while it may be that some physical quantities do not possess values independent of their measurement and are thus not 'real' in the sense in which 'realism' is typically defined, the properties that are associated with those quantities objectively exist. In fact they *must* exist, otherwise we could never properly construct measurement devices in the first place since we would have no way of guaranteeing that they would ever provide us with meaningful results. In short, our ability to make comprehensible measurements guarantees that at least some level of objective reality exists. I will have more to say on comprehensibility in a moment.

### 1.2.4 Local Realism

Famously, quantum mechanics seems to force us to choose between locality and realism. Consider the prototypical Bell-type experiment in which a source emits pairs of entangled particles in beams that propagate in opposite directions. Alice controls a measurement apparatus along one beam and Bob controls one along the other. They perform their measurements whenever a particle arrives from the source. As such, we assume that entangled pairs of particles will arrive simultaneously at both Alice's and Bob's measurement devices making it impossible for them to communicate their results classically, i.e. the events marking the simultaneous measurement of an entangled pair must be spacelike separated. Both are entirely free to choose the basis in which to measure the entangled property of a particular particle on their beam. Sometimes they may choose the same basis and sometimes they may not. However, as we know, when they happen to choose the same basis for a given entangled pair, they will necessarily find that their results are anti-correlated for the given observable in question. This presents a conundrum. As pointed out by Clauser and Shimony, "either the thesis of realism or that of locality must be abandoned" [11], p. 1883.

But we've made an assumption here that turns out to be incorrect and it is my entirely unscientific belief that this assumption is what leads many scientists and philosophers to choose to abandon realism rather than locality. That assumption is that causality logically requires locality. In fact it doesn't [52]. As Popescu and Rohrlich have argued, quantum mechanics actually reconciles causality with *non*locality [45]. The quantum correlations in Bell-type experiments cannot be exploited to transmit information superluminally [31]. As such, there's nothing that particularly favors locality over realism. Again, that's not to say that there aren't issues with realism, but, as I pointed out above, these issues do not rule out some level of objective reality.

## 1.2.5  Comprehensibility and Computability

I introduced the Principle of Comprehensibility above and I wish to briefly expand on this point in closing out this section. It could be argued that comprehensibility is the greatest argument in favor of an objective reality; the mere fact that the universe is comprehensible would seem to *require* an objective reality. Consider the following point made by Davies.

> Given the limitless variety of ways in which matter and energy can arrange themselves, almost all of which would be "random," the fact that the physical world is a coherent collection of mutually tolerant, quasi-stable entities is surely a key scientific fact in need of explanation [14], p. 61.

He goes on to say that unless one accepts the fact that the regularities that emerge from this apparent randomness "are in some way objectively real, one might as well stop doing science" [14], p. 62.[7] To bolster his argument, he addresses the issue of computability. As he notes, most mathematicians likely subscribe to the Church-Turing thesis which says that any function that is computable by a human being (ignoring resource limitations) must also be computable by a Turing machine (universal computer). In other words, if a mathematical problem is solvable then the Church-Turing thesis guarantees that a Turing machine can solve it. It was Deutsch who then pointed out that this must have implications for the physical world since any such machine would necessarily be physical and thus would obey the laws of physics [15]. In fact, as Deutsch notes, even the mere fact that *we* can compute

---

[7]It is perhaps ironic that the very first paper in the exact same collection as Davies' was Wheeler's famous 'it from bit' article in which he argued that there should be no immutable physical laws [61].

things as human beings (not to mention the several other species that have been shown to make simple computations) is because the laws of physics allow for it! As Davies observes, then, if the physical laws that govern the universe were considerably different than they are, then some operations that are computable would no longer be. To put it another way, if the laws of physics were subjective or malleable in some way, then something that is computable to one observer might not be computable to another observer. The fact that simple arithmetic is not only understood by humans but is also observed in other species, seems to be fairly convincing proof to the contrary.

To briefly summarize this section, I have laid out a generally broad argument in favor of an objective reality. I have specifically addressed several of the usual arguments against it and shown, in each case, that something objective persists. Indeed, there really is no point in doing science unless some level of objective reality exists. How else would one prove something? The concept of 'proof' relies on certain tacitly agreed-upon premises. For instance, for me to prove to you that $2 + 2 = 4$, I am relying on the fact that we both agree on the concepts '4' and '2' and the meaning of addition. If we can't agree on these basic elements of objective reality then there's no point in even discussing the matter. But even then, the mere fact that we can communicate to begin with seems to be fairly tangible evidence of something objective.

## 1.3   Mathematics and Computation

If the laws of physics dictate what is 'computable' then it is necessary for us to define 'computability'. Luckily, it is already well-defined within mathematical logic and theoretical computer science. In essence, computability is interpreted as a rough measure of the effectiveness of a solution to a given computational problem. As such it is very closely linked to whether or not an algorithm exists for solving that problem. There are several types of computational problems. For example, suppose we are given a positive integer and we are asked to determine whether this positive integer is prime. This is often referred to as a *decision* problem. On the other hand, given a quadratic equation and a series of values for the independent variable, solving for the corresponding values of the dependent variable constitutes a *functional* problem. These are just two of several types of computational problems.

If a computational problem is computable, then that means that an algorithm exists to solve it. In fact there may be more than one since each algorithm generally holds within a given formal *model* of computation, of which there are many. The Turing machine is just one of these formal models. It might be possible for a given problem to be computable in one model but not in another. Indeed, each model has its limitations. So if a problem is provably not computable in a given model, it still might be provably computable in some other model and ambiguous in some third model. All of this is meant to say that it is not always obvious whether or not a given problem is computable.

In examining the laws of physics and their relation to mathematics, it is necessary to wade a bit into the murky waters surrounding the nature of mathematics itself. Those who tend to study computability directly, notably theoretical computer scientists and some linguists, tend toward a formalist view of mathematics (at least according to Davies [14]—I have made no formal study of this claim). The formalist view interprets mathematical operations as nothing more than mappings. As such, formalism tends to view mathematics as 'invented.' On the other hand, physicists and mathematicians deeply interested in foundational issues tend (again according to Davies) toward mathematical Platonism which views mathematics as discoverable. This is an age-old debate that I have no hope of settling here. However, there are some important implications of each view for the nature of the physical universe and for how science itself works and so I will spend some time discussing each.

### 1.3.1 Mathematical Formalism and Representability

In some sense, the formalist view is true regardless of any ontological status that we might assign to the associated symbols. Mathematical operations *are* ultimately mappings as they all essentially take something and transform it into something else (see Fig. 1.2). Even decision, optimization, and search problems can be interpreted in this way. This is potentially what Wheeler was getting at when he posited that, at its core, the universe was nothing more than answers to 'yes-or-no' questions [61]. This view of mathematics fits quite naturally with an operator-based view of quantum mechanics and a transformation-based view of relativistic dynamics. Indeed, it fits quite well with the general notion of an observer in physics.

**Fig. 1.2** In some sense, mathematical operations are nothing more than mappings, transformations, or I/O operations. This is the essence of computability and representability

While the various computational models differ in their internal workings, they all essentially accomplish the same process in the end. This process is outlined in Fig. 1.2[8] which provides a fairly stark example of why one might be tempted to make the leap to Wheeler's view and reduce all of mathematics to answers to 'yes-or-no' questions. But that's quite a leap to make since it is not necessarily clear that all mathematical statements and operations can be reduced to binary outcomes. One might argue that the mere fact that mathematical statements and operations can be written on computers, which are binary machines, is proof that they can be reduced to Wheeler's ideal. But they can also be written down by human beings and it is not clear that we are necessarily binary. In fact, such an argument is really circular—the fact that computers give binary results can't be used to prove that they are inherently binary.

To understand this better, consider the operation shown in Fig. 1.2 to be an actual physical primitive or element from which we can construct a device that will display a *representable* number. The idea of a representable number is borrowed from floating-point arithmetic in computer programming and is different from Turing's notion of a computable number [54]. A representable number is simply a number that can, in some manner, be *exactly* representable by a 'machine.' So, for instance, any reasonably-sized whole number could be exactly represented by a collection of similar items equal to that whole number, e.g. four elephants represents the number four. Ideally, however, we would like a machine that is capable of exactly representing *any* number, however large or small. Thus, aside from the impracticalities of collecting elephants for this task, it makes far more sense to define this 'machine' in terms of the operational elements needed to represent a given number. To give a more concrete example, consider the following number written in both decimal and binary:

$$1197_{10} = 10010101101_2$$

Suppose that our primitive operational element, as symbolized by Fig. 1.2, has ten possible outcomes ('0','1','2','3', etc.). Then we could construct a decimal-based machine on which this number is exactly representable using just four of these elements. Conversely, if our primitive operational element only has two possible outcomes ('0' and '1'), then we would need eleven such elements to exactly represent this number. In this way our collection of operational elements acts essentially as a kind of *register* (as in a processor register in computing).

Now if our machine is decimal-based, then the simple, terminating fraction 1/10 is exactly representable. On the other hand, if our machine is binary, then we must represent the fraction as an expansion,

$$\frac{1}{10} \approx \frac{1}{16} + \frac{1}{32} + \frac{1}{256} + \cdots = \frac{1}{2^4} + \frac{1}{2^5} + \frac{1}{2^8} + \cdots$$

---

[8] I use this figure routinely when I teach both quantum mechanics and pure mathematics courses. I have been known to refer to it in the quantum context as a 'quantum meat grinder' since it resembles an old fashioned meat grinder. Independently, Chris Fuchs (who also contributed to this volume) has been known to do the same.

where each denominator in the expansion is represented by some collection of binary operations. As we can see, this expansion is non-terminating and so the number is not exactly representable on our binary machine unless the machine itself is infinitely large, i.e. includes an infinite number of operational elements. In floating-point arithmetic, the general rule is that a representable number is any rational number with a terminating expansion.

The most important difference between computable and representable numbers can be understood when one considers that irrational numbers such as $\pi$, $e$, and $\sqrt{2}$ are computable but not representable. That is to say, Turing takes it on faith that a sufficiently large or sufficiently robust machine can be developed to compute a sequence, no matter how large. Infinities do not necessarily pose problems for computability. Representability pragmatically asks what is possible given the type of machine. This has implications for the universe if we take Wheeler's claim at face value. If Wheeler is correct and the universe can ultimately be boiled down to binary operations then there are simple numbers that are not representable at the fundamental level if the universe is finite. The decimal system (indeed any base other than two) would be emergent in such a case.

Whether or not we take Wheeler's 'it from bit' at face value, the preceding argument is a perfect illustration of Davies' point. Imagine a toy universe in which Wheeler is correct and the most fundamental entity is a bit as represented by an operational element (Fig. 1.2) with two possible outcomes. Everything in this universe is constructed from bits and so, in a sense, it can be viewed as a kind of physical 'law' for this universe. In such a universe only certain numbers are representable at the fundamental level. On the other hand, consider a toy universe in which the most fundamental entity is a decimal number as represented by an operational element with ten possible outcomes. In such a universe, again only certain numbers are representable at the fundamental level. Some numbers representable in the binary universe are not representable in the decimal universe and vice versa. This is because the core physical 'law' describing how numbers are represented, is different in each case. If we were to apply the principle of relativity to either of these universes it would require that this core physical 'law' of representation would need to be relativistically invariant. Thus either 'it from bit' is universal, or it is simply wrong.

If we move to the domain of computable numbers, as with representable numbers, we find that there are certain things that are not computable, even approximately. In fact the full ordering relation on the computable real numbers is not even approximately computable nor is the equality relation between any two computable real numbers. As Deutsch has pointed out, this should have real, physical consequences meaning that there must be physical processes that correspond to non-computable mathematical descriptions. In fact such processes have been predicted; Geroch and Hartle give an example from quantum gravity involving non-computable sums over topologies in the calculations of certain path integrals [29].

Both representable numbers and computable numbers highlight a problem with mathematical formalism. In either case we are confronted with something that we know exists but that cannot be represented or computed. But this is not merely an abstraction according to Deutsch. It has profound implications for objective reality:

we have definite knowledge of something, i.e. we know what it should be, and yet we cannot construct a 'machine' to represent it or compute it. Perhaps the problem is that the universe simply can't be reduced to simple operations and we have no choice but to accept the Platonist argument.

### 1.3.2 Mathematical Platonism

In a sense, this debate over the nature of mathematics parallels the debate over the wavefunction in quantum mechanics with interpretations that range from ontic to epistemic. If mathematical structures are merely mappings, then mathematics could be interpreted as epistemic. In this way, the problem we just encountered with mathematical formalism could be addressed by simply saying that there are systemic limits to our knowledge of reality. If this is intellectually unsatisfying, one could turn to Platonism. To understand the force of such an argument, consider Tegmark's mathematical universe hypothesis [53].

**Definition 1.4** (*Mathematical universe hypothesis (MUH)*) Our external physical reality is a mathematical structure.

What Tegmark argues is not only that the universe *is mathematics* but that every possible set of initial conditions, physical constants, and sets of equations are equally real. Thus, in Tegmark's view, both our toy universes—binary and decimal—are equally real. It is no surprise, then, that Tegmark generally supports a multiverse model.[9]

In the Platonic view, mathematics is discovered rather than invented. While the symbols are merely representative, the underlying processes that define the mathematics are real. Platonism does not reduce all of mathematics to simple operations. Thus, while the formalist argument might leave one unsatisfied when encountering things that can be predicted but not constructed, the Platonist view simply accepts these things as real. Thus infinities are not a problem for Platonism. The question of actuality, i.e. of constructing a machine or performing a calculation, never arises.

One of the appealing aspects of Platonism is that mathematics, unlike physics, is entirely self-consistent even when considering Gödel's theorems. These theorems simply limit what we can *prove* about mathematics. They do not identify any internal inconsistencies. So while quantum mechanics and general relativity do not seem to work well together, no branch of mathematics proves especially problematic for any other branch of mathematics. This is perhaps not surprising given that mathematics is almost entirely built on formal logic (including Gödel's theorems).

Arguments in favor of a Platonist view of mathematics are certainly forceful when one considers simple arithmetic. If mathematics is invented rather than discovered,

---

[9]I should point out that it is not necessarily true that there need to be multiple universes as such. As Everett's biographer Peter Byrne pointed out, Everett argued that all instantiations of the wavefunction were real. He did not argue for a multiverse model. The multiverse interpretation of Everett's argument is due to DeWitt [9].

how is it humans are not the only species that can perform simple arithmetic? Many species, including the humble pigeon, display a basic understanding of not just counting but of ordinality, cardinality, and even some set theory [10, 49]. In fact, animals can even discern subtle differences in geometry. For example, considerable work on shape differentiation in dogs was performed in the very late 19th Century and very early 20th Century by Orbeli (who also studied their perception of color) and Shenger-Krestovnikova. Work by the latter found that dogs could differentiate an ellipse from a circle up to the point at which the ratio of the semi-major to semi-minor axis was 9:8 [44]. One could argue that this is nothing more than an ability to differentiate patterns, but then one could also argue that mathematics is nothing more than the systemization of patterns and symmetries.

The Platonist response to the problem of non-representability is to recognize that, to some extent, it is a problem of symbology. Take an infinite set as an example. We could never hope to write down all the elements of such a set on a piece of paper as it would take an infinite amount of time and an infinite amount of paper. However, we could assign a symbol, $\aleph$, that we understand fully represents this set and all its elements. Thus we have a way to finitely represent something that is actually infinite simply by providing it with a symbol. If all of mathematics is equally real, then perhaps the physically 'real' aspect of the infinite set is the symbol we assign to it which is, itself, finite. The infinite number of elements contained in the set are merely the *meaning* we assign to that particular mathematical symbol. Meaning is an abstraction and thus does not necessarily possess a physical ontology.

Nevertheless, the Platonist view has its drawbacks. In Tegmark's argument, our *physical* reality is nothing more than mathematics meaning that Tegmark and Wheeler are closer in their view than they might at first appear. In both views, the fundamental primitives of physical reality are a kind of abstraction. But as I argued in Sect. 1.2, physical reality has a clear set of basic rules that it must obey in order for it to be comprehensible. What, then, am I to make of the fact that there are things to which I can assign an abstract meaning but that I know cannot physically exist? We might call this the 'unicorn problem.' How is it that we can draw pictures or make models of unicorns and yet we know that such animals do not exist?[10] How is it that we can conceive of small spacecraft hurtling around outer space with warp drives and a simulated gravity field equal to Earth's actual gravity field, and yet we know such things defy the laws of physics?

### 1.3.3  Mathematics as Language

Several approaches to this problem present themselves. One option is to view mathematics simply as a language with strictly enforceable rules (unlike English!). Indeed, while it would prove tedious and virtually impossible to read, one could simply rewrite mathematical problems in a spoken language. Early treatises on mathematics

---

[10]Recent discoveries of the 'Siberian unicorn' notwithstanding.

and physics were written this way since the formalism had not been invented yet (e.g. Newton's *Principia*). Now, one could argue that mathematics is not the same as its formalism. But one could also argue that the non-symbolic aspects of mathematics, i.e. the underlying 'rules' that dictate how the symbols can be combined, are really just a manifestation of physical laws. In fact one could argue that they arise from the fundamental principles and axioms that form the foundations of physics. Then, instead of thinking of physics as being built upon the foundation laid by mathematics, we might think the opposite: mathematics arises from physics just as any other language arises from its 'environment.' As such, the limits to mathematics are only in our imaginations and in the formal rules of self-consistency that we impose on it. In a sense, mathematics can be viewed as a *formal language*. The concept of a formal language was first of introduced by Frege in 1879 [27] and was more fully developed by Chomsky, Schützenberger, Kleene (whose doctoral advisor was Alonzo Church), and others in the 1950s (see, for example, [43]). The exact nature of the relationship between formal language and mathematics is still being studied (see, for example, [7, 28]). But, to my knowledge, there has been no attempt to describe mathematics as being emergent from the underlying physics in the sense that without the actual underlying physical processes, mathematics would have no meaning (though constructor theory, as outlined in Marletto's chapter in this volume, suggests such a relationship). Such an endeavor is a daunting task and I make no claim to having accomplished it. Nevertheless, a few examples with toy universe models can illustrate why this might be an alluring alternative.

Earlier I introduced the Principle of Comprehensibility and will now employ it in analyzing several toy universe models. Recall that this principle holds that the nature of a physical system under investigation is bounded by the methods of that investigation. The key here, is that there must be a method of investigation to begin with. Otherwise we're not dealing with a physical system.

So first let's consider a toy universe that is entirely empty. A cosmologist or general relativist might immediately refer to such a universe as a de Sitter universe. A de Sitter universe, however, is not truly empty, per se, since it may possess dark energy or the inflaton field which are 'things' in the sense that they are distinct from the universe to which they are associated. A de Sitter universe might be better referred to as a vacuum universe or matterless universe where we understand matter to be something that has mass. Our empty toy universe literally has nothing in it—no matter, no mass, no fields, no energy, nothing. As such, it is almost meaningless to even speak of such a universe. Indeed, comprehensibility tells us that such a universe would not even be physical since no method of investigation exists by which to analyze it.

To that end, allow me to introduce the term *primitive* to mean any 'thing' associated with (i.e. 'in') a universe that cannot be subdivided and, when considered alone, has only a single configuration (i.e. a primitive has zero statistical entropy when considered without any context). Primitives are the fundamental buildings blocks of anything within a universe and can include fields, fundamental particles (which are really just quantized fields), etc. Dark energy and the inflaton field would be primitives in a de Sitter universe where the context that allows them to have non-zero statistical entropy would arise from the imposition of a metric.

Now consider a universe that contains a single primitive. We might rush once again to judgment and assume that this is a de Sitter universe or some other such thing. In other words, we might be tempted to ascribe to this universe concepts such as spacetime coordinates, metric tensors for describing the lone primitive, or we may even be tempted to give some physical nature to the primitive (e.g. dark energy, inflaton field). But this is an inherent bias that comes from our knowledge of *our* universe. There is absolutely no reason to assume that this toy universe even follows the same physical laws as the one we inhabit. For all we know, general relativity might not even apply. Since the single primitive has zero statistical entropy when considered alone, this universe also does not pass the test of comprehensibility. In such a universe, simple arithmetic has no meaning—addition, subtraction, multiplication, and division are entirely meaningless when only a single primitive exists. Neither are more abstract concepts such as groups, functions, etc. In fact space itself is difficult to interpret in a universe with a single primitive regardless of whether one views it, as Leibniz did, as a purely relational concept, or whether one takes Kant's view of space as part of a systematic framework by which we can interpret the world. Leibniz's interpretation fails for obvious reasons (there is nothing to relate the primitive to) and Kant's interpretation fails because there is nothing to interpret.

A question naturally arises from this: why *can't* we simply assume that this universe obeys the same laws of physics as ours? The answer is that we certainly *could*. Indeed, most studies of 'alternate' universes, i.e. those with different physical laws, really only take the laws of our universe and perturb them slightly. Most often, such studies simply change some physical constant or constants (e.g. the three light quark masses [34]). However, there again is no a priori reason to assume that if other universes exist they must only be slight variations on ours. Granted, there would have to be some consistency across any connected multiverse, but the point is that nothing rules out universes that are vastly different.

Returning to our toy universe, let us introduce a second primitive. We might now be tempted to introduce all manner of quantities and mathematical structures. What I'm interested in arguing here is that certain mathematical structures inherently depend on a specific physical context. That's not to say that we can't abstract those concepts, but the point is that they only have meaning in the first place as a result of some physical context. Consider simple arithmetic. In a toy universe with two primitives, basic arithmetic still does not arise without some way in which these primitives may combine. For example, consider two possible variations on a two-primitive universe. In the first, the primitives may be able to combine to form a single, non-primitive (i.e. composite) entity. In the second, no mechanism exists for combination. Clearly, there are mathematical structures that exist in the first universe that have no meaning in the second since, in the first, we have a notion of combination while in the second we do not.

The counter-argument to this is to say that just because a particular physical structure does not exist in a given universe does not necessarily mean that the mathematics
that might describe such a structure doesn't exist. After all, there would seem to be
mathematical structures in our own universe that do not have physical corollaries,
most obviously the concept of infinity. As far as we know, it is impossible to actually
measure an infinite amount of anything and yet the concept exists.[11] Note that, as
a concept, infinity must be logically preceded by the concept of cardinality. Infinity
makes no sense unless one understands, even at a very basic level, the concept of cardinality, i.e. of 'number' or counting. The idea of cardinality ('number') arose from
a natural need by humans (as well as a few other species) to describe or understand
the universe. In other words, a physical structure led to a mathematical structure.
Additional physical structures certainly drove the development of infinity as well,
since it, too, arose out of a need to describe certain concepts. But it is still, to a large
extent, an abstraction of the more basic concept of cardinality (counting). This is
even true when we consider structures such as Cantor sets and Dedekind cuts. After
all, dividing and sub-dividing something increases, in some sense, the number of
partially independent 'things' one has.

It is, therefore, my contention that mathematics is best viewed as a language
that formally describes the physical structures inherent in the universe. A different
universe might have entirely different mathematics.

In summary, as Eddington has noted, we too often ignore the initial insight (usually
physical or logical) that led us to a particular mathematical deduction in the first
place [20]. It is this initial insight that ultimately gives birth to the mathematical
deduction. The view of mathematics as a language holds that while the initial insight
may form some kind of mathematical structure in the sense of a definitive pattern that
is describable by mathematics (the language), mathematics itself is a *language* and
can describe anything within the bounds of its own rules of self-consistency even if
there is no physical corollary. This strikes a balance between the traditional formalist
and Platonist views and helps to explain the 'unicorn problem' in which there are
things that can be proposed that seem to violate the laws of physics.

## 1.4 Physical Laws and the Nature of Scientific Discovery

The ideas outlined in Sects. 1.2 and 1.3 are both key to understanding how science,
and particularly physics, works. I have argued that certain core physical laws form
the basis of an objective reality. As Eddington noted, science is ultimately just a
"rational correlation of experience" [20] (p. 184) which underscores Davies' point
that there must be something objective about the universe, otherwise we would never

---

[11]There certainly is some debate about this, particularly if we equate continuity with infinity, but
if we take 'counting' as fundamental, then there simply is no way that we know of to 'count' an
infinite number of 'countable' things in a real, physical sense.

be able to correlate our experiences and might as well just give up. But there are two critically misunderstood points about the nature of physical laws that urgently need to be addressed.

### 1.4.1   Physical Laws as Relational Structures

Theories have been proposed over the years that have been interpreted as suggesting that the laws of physics could be variable. In fact most such theories actually propose that it is the values of the *constants* in the equations of those laws that vary, not the equations themselves. Regardless, these theories are often sold to the public (and sometimes to fellow scientists) as suggesting radical changes to physics when the truth may not be as jarring.

Consider theories that propose that the speed of light in a vacuum, $c$, might vary. Einstein himself spent close to a decade wrestling with this idea [23–25]. While he concluded that the speed of light was a constant in *most* cases, it is not clear from his own writings whether or not he felt it was constant in a changing gravitational field. The most widely accepted current view, argued by Bergmann, assumes that the *vector* describing the velocity of light can change in a changing gravitational field but that the actual *speed* (a scalar) cannot [4]. Nevertheless, many other theories based on a varying speed of light (VSL) have been proposed, particularly in the past few decades (see [38] and references therein). It is not entirely clear what the main objection to VSL theories actually is.

On the one hand, some objections appear to be centered around the fact that the speed of light in a vacuum appears as a constant in all our best theories and, because those theories appear to be correct and empirically supported, we are justified in our assumption that it is a constant. But of course time is treated as absolute in Newton's laws and the best measurements for more than two centuries seemed to confirm this even though we now know it only to be an approximation that holds for $v \ll c$ and low energies. So this is hardly a valid argument against VSL theories.

On the other hand, one could object to VSL theories based on the argument that $c$ must be a constant and a maximum in every inertial reference frame in order to preserve causality. But, in fact, this is not entirely true. Not only does a spatially varying speed of light *not* contradict general relativity, it is, in fact, inherent within the theory itself. For example, for a gravitational potential $\phi(r)$, the speed of an individual photon can be written as $|u| = 1 + 2\phi(r) + \mathcal{O}(v^3)$ [56, 62]. It is entirely within the realm of possibility that our existing theories are what one would get when $c(r, t) \approx$ CONSTANT just as Newton's laws are what one gets when $v \ll c$. In addition, relativity merely requires that $c$ is a *local* maximum and it never specifies what *value* $c$ must have. Indeed, the values of *all* physical constants are empirically derived and most are not predicted by theory precisely because values of physical constants depend on systems of units, e.g. the value of $c$ in SI units is not the same as the value of $c$ in Imperial units.

Observations such as these are what drove many physicists, including Eddington [21], to attempt to develop unit-free theories of physics. Indeed, if we maintain that $c$ must be a maximum in every inertial reference frame and then we switch to a 'natural' system of units in which $c = 1$, all speeds become unitless ratios which maintains the primacy of $c$ *within* any given reference frame. It is even relatively simple to show that the relativistic velocity transformations *between* frames still hold. The point is that the nature of the underlying physical law here is a *ratio*.

The fact is, physical laws are *relational* structures that allow for the comparison of physical quantities. It is typically some mathematical or logical structure that defines a physical law, never any particular numerical quantity. So proposals that call for variations to the values of physical constants are not necessarily in violation of any physical laws so long as the various ratios and relations *between* quantities remain unchanged. This is because it is these ratios and relations, i.e. the mathematical and logical structures, that define the physical laws and *not* the numerical values of any associated constants.

## 1.4.2   The Evolution of Physical Laws

Plenty of 'new' theories have been proposed over the centuries that *have* called for changes to the mathematical and logical structure of physical laws. Many people, including many physicists, often view this as how science progresses. It is assumed that with every new paradigm shift (in the Kuhnian sense [36]), we debunk our previous theories and adopt new ones. To adherents of this approach, science is mostly a process rather than an accumulation of knowledge. I have no desire to tackle the intricacies of this debate. I have already argued that there must be an underlying objective reality. As such, not everything is relative and there must be some 'truths' that are uncovered by science. I have argued above that these underlying 'truths' are the physical laws that govern the universe. The main objection to this line of argument is that paradigm shifts cause a wholesale replacement of outdated physical laws with new ones. For example, it is popular to say that Einstein's development of relativity proved Newtonian mechanics was wrong. But this view of paradigm shifts in science is perhaps the most egregious misunderstanding of physics. While it may be true that such things happen in other fields, as Weinberg has pointed out, since the caloric theory of heat, no widely accepted physical theory has ever been proven to be wholly and unequivocally wrong [57, 58]. Rather each new theory simply shows that the previous theory was merely an approximation, valid under limiting conditions. This is a very important and deeply misunderstood point that bears a more thorough analysis.

The first question we must ask ourselves, when confronted with a theory that is said to have been proven to be 'wrong,' is what about the theory was 'wrong?' Since the advent of modern science (generally accepted to correspond to the time of

Galileo and Kepler), it (and particularly the physical sciences) has followed a specific process (the scientific method). As Feynman has noted, the absolute arbiter of any theory is—*and must be*—empirical evidence [26]. Regardless of how beautiful any theory may be, if it contradicts empirical evidence, it is wrong. Of course, there's a bit more to it than that, but at the end of the day if your theory predicts that the sky will be green but your observation shows that the sky is blue, then your theory must be wrong. According to Feynman, then, a theory's 'truth' is determined by how well it compares to the physical reality that we measure and experience. In this sense, Newtonian mechanics is not wrong. In fact we rely on the fact that it is *right*, at least within the space of our everyday existence, since it is used by engineers to build bridges, roads, automobiles, buildings, airplanes, and a host of other things. Our lives literally depend on it to be right.

What Einstein actually showed was that Newtonian mechanics was an approximation of a more general theory (relativity). One obtains all of Newtonian mechanics in cases where $v \ll c$, energies are generally low, and gravitational fields are weak; Newtonian gravity and Newtonian dynamics are merely a limiting case of general relativity (for a complete proof see Chaps. 7 and 8 in [50]). In fact this *must* be the case since Newtonian mechanics and Newtonian dynamics have empirical support in those regimes.

It is difficult to argue with mathematically rigorous arguments and even more so when those arguments match experiment. Perhaps people who continue to state that Newton was wrong, do so because his three 'laws' do not necessarily generalize very well to curved spacetime (though this is a somewhat debatable point). Laws, they may say, should be infallible truths about the universe. But that really just means that we have historically mis-labeled them as 'laws.' The fact remains that they are a limiting case of general relativity.

Or perhaps people object to the fact that Newton viewed time as absolute. While the 'truth' may be that time is *not* absolute, it is *approximately* absolute in low-energy, low-speed, and low-gravity regimes. Or, to put it another way, in the case in which gravitational fields are weak, general relativity is approximated by special relativity. In the case in which gravitational fields are weak and energies and speeds are low, general relativity is approximated by Newtonian mechanics. Yet we never seem to say that special relativity is 'wrong'. We simply say that it is a limiting case of a more general theory (relativistic quantum mechanics and quantum field theory).

I have found it interesting to compare the case of Newtonian mechanics to that of electromagnetism. Electromagnetism is fully described by Maxwell's equations and the Lorentz force 'law.' Because they are relativistically invariant, there seems to be an assumption that they are 'correct.' But they only describe classical electromagnetic interactions. How is that limitation any different than the limitation of Newtonian mechanics to low-energy, low-speed, and low-gravity regimes? Every theory has its limitations. Certainly some are more broad than others, but all are eventually inadequate to completely describe the universe. If the argument against Newtonian mechanics is based on how Newton, himself, interpreted the laws, then why not hold electromagnetism to the same standard? Maxwell's equations were developed with the notion of the aether in mind. Maxwell even proposed models for devices aimed at detecting such an aether with FitzGerald actually constructing one of those models. In fact, electromagnetism was developed under the assumption that Newtonian mechanics was universally correct. It was only *later* discovered that Maxwell's equations and the Lorentz force 'law' were relativistically invariant.

The point is that all theories have contexts within which they are developed and all theories are limiting cases of some larger, perhaps unknown, theory (they must be, since we do not have a theory of everything as of yet). Sometimes we find that those contexts or the preliminary assumptions we made in order to derive the theory are incorrect but that the theory still does an admirable job of modelling the experimental evidence. This is largely because all modern physical theories are built upon two main pillars: empirical evidence and a corresponding rigorous *mathematical* description. Our interpretation of both may change, but the evidence and the associated math are never fully supplanted.

### 1.4.3   Further Considerations

While empirical evidence and mathematical rigor are the two pillars that support modern physics, there are many other criteria that play a role in how it functions. In talking about science in general, Kuhn outlined five criteria that scientists use to determine the general validity of a theory [35]:

1. **Accuracy**: This is just the empirical evidence argument that I have already discussed in depth.
2. **Consistency**: Theories should be internally self-consistent as well as externally consistent with other widely accepted theories. Generally, adherence to mathematical and logical rigor as well as empirical accuracy, will ensure consistency in most cases.
3. **Broadness of scope**: We aim to find theories that describe as broad a swath of physical reality as possible. The reason for this is best understood if we consider an example. Suppose we have two theories such that theory $A$ describes phenomenon $P$ while theory $B$ describes phenomenon $Q$. Now suppose that we discover a new phenomenon that has some characteristics of $P$ and some characteristics of $Q$. Which theory do we use to explain this new phenomenon? The best case scenario is to find a theory that describes $P$, $Q$, and any mixture of $P$ and $Q$. If theories $A$ and $B$ are already accurate and consistent, our best hope is to find a third theory for which $A$ and $B$ are limiting cases.
4. **Simplicity**: Suppose I have a phenomenon $P$ that is fully described by theory $A$. I could propose a second theory, $A +$ aliens, which also fully describes $P$ (because it inherently includes $A$), but what would be the point? I only need $A$ to describe $P$. I don't need the additional hypothesis of aliens. This, of course, is just Occam's razor.

5. **Fruitfulness**: A good theory should be predictive of new phenomena or new relationships between existing phenomena. One only needs to look to astrology (which, of course, is *not* a science) to see 'theories' that are terrible at prediction.

Kuhn goes on to argue that whichever theory ends up being most widely accepted is dependent on subjective arguments among scientists. But, I am not interested here in Kuhn's additional remarks since I have already addressed subjectivity in Sect. 1.2. What I am interested in pointing out is that, beyond the rigorous mathematics and empirical support, theories do generally require some interpretation. After all, every symbol in every mathematical equation of every physical theory has some physical interpretation or meaning. Any physical theory that contains a mathematical symbol without physical meaning, even if indirect, can not lay claim to being a complete physical theory. Even complex numbers possesses meaning, albeit indirect, in that they allow for certain physical combinations in quantum mechanics that do not make sense without them. So while they are not directly measurable, they are nevertheless required for the measurements within quantum mechanics to be consistent. As such they possess some indirect physical meaning.

What I propose is that it is this *meaning* that is subject to change and updating as science progresses. So, for instance, our understanding of the familiar $t$ in all our physical equations has evolved considerably over the past four centuries. We now know that, in some cases, $t$ (coordinate time) may not appear to have the same value for two different observers. At worst, we may have found that our existing equations containing $t$ were a limiting case of something more general that might or might not involve $t$. But nowhere in any of our theories was it necessary for us to completely abandon $t$ or the equations that contained it. We simply have increased our understanding of the physical meaning behind the symbol and have found that, sometimes, our existing equations are limited in their scope. This, of course, is precisely how language works and underscores many of the points I made in Sect. 1.3: meaning evolves and sometimes necessitates the creation of new words or new combinations of existing words, but it never means that the old way of saying something is wrong. It may not be fully useful or understood in a particular setting, but it isn't wrong (in the sense that the syntax isn't wrong).

If all of what I have argued to this point smacks a bit of logical positivism, I would encourage the reader to withhold final judgment until the end. The aim of this section has been to bring the arguments of Sects. 1.2 and 1.3 together by arguing that the physical laws that constitute objective reality are understood through mathematical structures that are relational in nature. Advances in physics find greater generalizations of these structures and can alter our interpretation of the elements of those structures, but the basic relational structures themselves always remain valid within the context in which they were originally developed. A visual representation of this process is shown in Fig. 1.3. In the next section I address the limits of the box depicted in the figure.

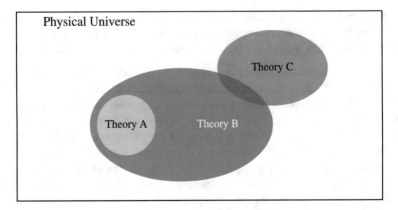

**Fig. 1.3** Theories describe different parts of the physical universe. In some cases there is partial or even complete overlap. In the figure, theory B fully describes all of the phenomena described by Theory A as well as quite a bit more. Assuming all the theories are mathematically consistent, Theory A should be a limiting case of Theory B. On the other hand, Theory B and Theory C both describe certain common phenomenon represented in the figure by their overlap, but neither is a limiting case of the other. What physicists desire is a single theory the describes the entire *rectangle* representing the physical universe

## 1.5 Limits on Knowledge

I have endeavored to make the contents of Sects. 1.2 and 1.4 as free from interpretation as possible. In other words, I am simply attempting to lay bare how science, and most particularly physics, actually works by laying forth the *facts*. I have advanced my own interpretation of mathematics in Sect. 1.3, but this should have no bearing on the truth of what I have to say about physics. Nevertheless, it is sure to see its share of criticism, some of which may be justified (after all, nothing is ever truly perfect). It might give the reader the impression that I support a radical notion of logical positivism or some such philosophy. Frankly I am generally agnostic on such things. But to assuage the reader who might be tempted to read too much into my previous remarks, in this section I outline the limits of science in a similar manner, i.e. I simply state the facts as I see them.

### 1.5.1 Classical Measurement Limits

The foundation of physical science is measurement. But what is measurement? Quantum physicists are used to viewing measurement as an active process by which the measurement changes the state of the system. Even classical measurements are in some sense 'interactive.' Take a voltage measurement, for example. Measuring a voltage between two nodes requires the creation of a high impedance path between those nodes. Without such a path, the voltage between the nodes cannot be measured.

Likewise, if the path were to suddenly switch from high impedance to low impedance (i.e. develop a 'short') it is entirely possible to fundamentally alter the system being measured, at least within the bounds of comprehensibility as outlined above.

On the other hand, to an observational astronomer, there is never the sense of direct interaction with the object being observed. Yet this is still a measurement. After all, in order to fully explain what we observe we need to actually record physical quantities. So, for example, we can make measurements of luminosity, spectrum, and relative motion. This is what separates an observational astronomer from someone who is just casually admiring the stars.

I will thus use the word 'measurement' in both an active and a passive sense and take it to mean the recording of the value of some physical quantity. This 'value' does not necessarily need to be numerical. Direction (e.g. of a vector) can be a non-numerical quantity. Obviously, in order to record that physical quantity the person or machine that is doing the recording has to either actively interact with the system being measured or passively observe it. The limits of most of our passive observations tend to be active in the end as the following example demonstrates.

Suppose we are interested in observing the position of a distant star relative to some arbitrarily defined coordinate system. We can get a rough estimate of the location of the star simply by looking at it and plotting its coordinates. The coordinates can be obtained by observing it relative to some other star that has known coordinates in our system. But the accuracy of the coordinates that we assign to the star will depend on how precise our observing instrument was. Perhaps we simply used our naked eye and then held up our thumb to roughly estimate how far apart the two stars are in the night sky. We could get a vastly more accurate location if we use a camera to get a high resolution image to which we can assign coordinates for every pixel—the higher the resolution, the more accurate the positioning. This positioning relies on our ability to mark precise locations on our image which is why even an analog image would need to be superimposed on a discrete grid of points in the end, i.e. at some point we have to assign an actual coordinate to each point in the image. Since it seems we can't escape discreteness in this example, we will have to eventually deal with individual photons. And while it is true that the observation itself will never change the state of the star, it *could* change the state of the photon carrying the information *about* the star.

We can see some of these issues more clearly through a slightly different example first suggested in [17]. Consider a rigid (classical) object moving at a constant speed. Suppose that we wish to measure this speed to as great an accuracy as physically possible, i.e. we want a truly *instantaneous* speed. Instantaneous speeds are well-defined mathematically, but are they physically measurable? We are assuming that we are external to the object itself and so the only way to really make such a measurement is to bounce light off of it. Common radar guns tend to work in the microwave regime and measure the Doppler shifts of the microwave signals bouncing off the object, i.e. as the object moves the wavelength of the signal changes and a phase shift arises between the wavelength measured at an instant $t_1$ and a later instant $t_2$. The light is assumed to be purely classical and continuous (I will discuss the quantum case in a moment). An instantaneous speed is defined as

$$v = \frac{dx}{dt} := \lim_{\Delta t \to 0} \frac{\Delta x}{\Delta t}$$

and so the key is to minimize $\Delta t$. Of course, that also implies we are minimizing $\Delta x$ as well since the object is moving at a constant speed. Recall that we are measuring this speed using *classical* light which treats light as a wave which is an *inherently non-local phenomenon*. In order to describe something as a wave it either must have some spatial extent (in order to define a wavelength) or some temporal extent (in order to define a period). Thus in order to make an accurate measurement of the Doppler shift of the light as $\Delta t \to 0$ (which requires also reducing $\Delta x$) we would need to employ light of increasingly shorter wavelength and frequency. Specifically, in the limit as $\Delta t \to 0$ we would require that $\Delta f \to 0$ and $\Delta \lambda \to 0$ for the classical light. This begs the question, can we construct a purely classical device to measure arbitrarily small wavelengths or arbitrarily high frequencies?

One method that suggests itself is to find some way to *infer* the overall properties of the wave from a measurement on an increasingly small portion of it. Consider the following measure known as the 'local wavelength' [41],

$$[\lambda(x)]^2 = -\frac{4\pi^2 f(x)}{f''(x)}$$

where $f(x)$ is the function describing the wave.[12] The local wavelength is a measure of the curvature of a particular wave at point $x$, i.e. it is theoretically an 'instantaneous' curvature for the wave. The value and sign of the local wavelength will give not just the direction of the curvature at a given point, but also will give a measure of that curvature, i.e. whether it is steep or gradual. The problem is that measuring it requires the ability to measure $f(x)$ and $f''(x)$ to arbitrary accuracy which is the problem we set out to solve in the first place. The fact is, in order to measure an instantaneous speed, we require the ability to measure a non-local phenomenon (i.e. the wave) perfectly locally which is simply impossible. To put it another way, despite the mathematically rigorous proof of infinitesimals, there remains no practical way to measure a given quantity to arbitrary accuracy if that quantity is a function of a *change* in some other quantity. This is because the definition of 'change' implies a duality by definition—one must have *two* of something in order to measure a 'change.'

It is clear, then, that at least some classical measurements are fundamentally limited in their accuracy. This is not necessarily true for *all* classical measurements. After all, if a given measurement merely represents counting macroscopic objects, there is no fundamental limit on accuracy. For example, we could presumably create a machine to count the number of marbles in a box to perfect precision. But the point

---

[12]I hesitate to use the term 'wavefunction' here because that might imply the quantum wavefunction. I am still working in the purely classical regime.

is that there are fundamental limits to *some* measurements that have nothing to do with quantum physics. These limits highlight one of the problems with mathematical Platonism: instantaneous change is mathematically well-defined but virtually impossible to measure to arbitrary accuracy. This is not a limit that can be overcome by advances in technology. At some point we must ask ourselves if this means that concepts beyond such a limit have any physical meaning.

### 1.5.2  Quantum Measurement Limits

Brukner and Zeilinger have argued that there is no such thing as a continuum [8]. In that case there is no such thing as classical light and thus it is obvious why we can never measure a truly instantaneous speed: since the wavelength of the quantized light is proportional to its energy, the time-energy uncertainty relation will prevent us from determining the change in energy of the photon and thus the phase shift to arbitrary accuracy as $\Delta t \to 0$. In other words, the accuracy of the measurement will be bounded by the *Heisenberg limit*.

It may be helpful to understand the Heisenberg limit in terms of interferometric systems and the *Fisher information*, which is a way of measuring the amount of information that a random observable carries about some unknown parameter of a distribution that *models* that observable. In standard quantum metrology, if we have a particular procedure that we repeat $T$ times and the Fisher information remains constant throughout then the mean square error scales as $\delta\phi \geq 1/\sqrt{T}$. This is sometimes called the *standard quantum limit* or the *shot-noise limit*. If the Fisher information is a function of the resources used, then the mean square error scales as $\delta\phi \geq 1/N$ where $N$ is the number of resources that are used. This latter scaling is the Heisenberg limit. So for example, in an interferometer the variable of interest is the phase difference $\Delta\Phi$ between the waves in the arms of the interferometer. We can view the photons as the resource in this case and so the Heisenberg limit is $\Delta\Phi = 1/N$ (notice that, if this limit holds, it prevents us from using such an interferometer to improve the accuracy of our Doppler shift measurements of instantaneous speed). At least one experiment claims to have circumvented this limit [42] though there is some disagreement concerning the scaling resource used in this experiment [30]. In fact, it was argued, at roughly the same time, that the Heisenberg limit is fundamentally unbeatable [64]. A deeper clarification of the meaning of the Heisenberg limit as well as a more detailed proof of it was subsequently given in [65]. Nevertheless, there have been proposals that have suggested different scalings for the mean square error [5, 47].

At any rate, the use of the term 'standard quantum limit' in interferometric systems is a bit misleading since there are broader limits related to both the traditional uncertainty relations, which prohibit the simultaneous measurement of pairs of complementary observables, as well as limits imposed by contextuality and entanglement, e.g. Tsirelson's bound which is an upper limit on the quantum correlation of distant events. Beating some of these limits would have profound implications beyond quantum mechanics. For example, beating the uncertainty relations for a pair of complementary observables would pose problems for both quantum mechanics as well as classical wave mechanics since one can derive uncertainty-like relations for classical waves. In fact the two differ only by Planck's constant [6]. It is highly unlikely for such a basic part of something as fundamental as wave behavior to be entirely wrong. Of course one can arrive at the uncertainty relations by an entirely different route that is almost purely mathematical. Pure mathematics (done correctly) is never 'wrong', though as Gödel's theorems show, it may be 'right' but not provably so. Either way, it is highly unlikely that the uncertainty relations can be circumvented, at least directly, since it would suggest deep flaws to both fundamental physical phenomena as well as possibly mathematics.

The key here is that it really is *phenomena* that produce these limitations. After all, as I argued in Sect. 1.3, physical phenomena are what guide the development of the mathematics that make up the formal portion of any theory. It is my contention that only phenomena that are observable, either directly or indirectly, have any meaning in science. Once we stray from something that can be physically provable, we are no longer doing physical science. The best we can say is that we're imposing speculative meaning onto pure mathematics. We're like an author who writes about unicorns. There's nothing wrong with writing about unicorns, but it's disingenuous to claim they exist simply because we have a language that allows for their description. Of course, the question that this raises is, how do we know if a particular speculative phenomenon is a unicorn or not? How do we know if a bit of predictive mathematics will eventually yield experimental results or not? After all it took a century for gravitational waves to be experimentally proven to exist. If there is one thing for certain, physics cannot describe something that is non-physical. In fact it can't describe anything outside of our universe—or can it?

### 1.5.3 The Physical Universe

Colloquially, a universe is defined as the totality of everything that exists [2]. The problems with this definition are numerous. First, it is inherently ambiguous in regard to both 'totality' and 'existence.' Second, it is not clear how a universe would be defined within the context of any theory that admits multiple universes, particularly in such a way that they could be distinguished in some meaningful way. In the recently proposed many interacting worlds (MIW) theory, for example, the 'worlds' all exist in a single configuration space and it is their relative 'closeness' within this space that determines the nature of any interaction between them [32]. So it is not necessarily clear that each 'universe' possesses a completely independent ontology. The nature of what we mean by a universe in such instances remains largely unsettled [3, 55].

Other notions of a 'universe' exist, of course. Physical cosmologists typically define a universe topologically as some kind of spacetime manifold, but this presents at least two problems. First, it assumes that the manifold is *real* and not merely a mathematical abstraction (see the discussion in Sect. 1.3). Second it is not clear how emergent spacetimes fit into this description (examples of emergent spacetimes include [33, 51, 59]). One could also attempt to define the universe based on a wavefunction of some kind [40], e.g. as a solution to the Wheeler-DeWitt equation, $H |\psi\rangle = 0$, but as long as the ontological status of the wavefunction remains unresolved [37, 46], this would seem to be a less-than-ideal way to define a universe. Operationally we could define a universe as the totality of all that can be *measured*. A radical take on this definition is Wheeler's participatory universe which implies that only things that can be measured can exist [60].[13] This view is surprisingly similar to that taken by Eddington who defined the physical universe to be the "theme of a specified body of knowledge" [20] (p. 3), specifically *physical* knowledge. As he later explains,

> The physical universe is the world which physical knowledge is formulated to describe; and there is no difference between the physical universe and the universe of physics [20], p. 159.

Eddington was a self-described *selective subjectivist* but it would be wrong to conclude that he did not acknowledge an objective reality. Indeed, he clarifies this point later in the same work and develops the details of his 'structuralist' approach to reality (existence) both there and in [19]. At any rate, Eddington's ultimate point is that the universe *is* the entirety of physical knowledge, both actual as well as *possible*; it includes both what can be directly measured and what can be indirectly *inferred*. Specifically it is defined as consisting of physical knowledge which consists of assertions of what has been or what *would be* the result of a specified measurement [20], p. 10. Wheeler, by contrast, might have claimed that anything that had not yet been measured would have no ontological meaning. So, for example, Eddington might view a single electron in a distant galaxy as being a part of the universe because we can infer its existence (if a distant galaxy exists, it must contain electrons even if we cannot directly measure them) whereas Wheeler would not be as generous.

Wrestling over the definition of a physical universe seems to be a Sisyphean task. One of the reasons for this could be that we are a *part* of one. Imagine being confined in one room of a house with no windows but with television monitors showing glimpses of some of the other rooms. You would have no idea what the exterior of the house looked like and, more importantly perhaps, you would have no idea if there were any other houses nearby or if your own house itself was actually infinite in extent. This is precisely the position we find ourselves in with regard to the universe we inhabit. We can describe our own 'room' with tremendous accuracy. We can draw some inferences with varying probabilities about what the rest of the house might be like or how it was constructed. It is unlikely, though not strictly impossible, that we would be able to infer anything about other homes or the full extent of our own home. But unless we can escape our room, we'll forever be limited in our knowledge.

---

[13]This is *not* the same thing as saying that only measurable things have meaning. Things can lack physical meaning because they are not reliably measurable and yet we know with certainty that they exist in some regard. Our own consciousness is an example of this, at least until we have a better understanding of it.

This actually suggests a physical correlation to Gödel's incompleteness theorems in the sense that in order to fully describe the universe we would need to step out of it. The fact is, there are things that are unknowable. Not only is it sheer hubris to think otherwise, it is logically inconsistent.

### 1.5.4   Interpretation and Extrapolation

There is one final point I wish to make in regard to the limits of science and particularly physics. We too often over-interpret or over-extrapolate from our results (and sometimes we under-interpret or under-extrapolate). This may be a result of the human desire for narrative or it may be driven in part by our seemingly increasing need to sensationalize everything. But it is a trap that we all too often fall into. As an example of over-interpretation, consider that most cosmologists and general relativists view spacetime as a manifold whose status is purely ontic. That is, they believe wholeheartedly that, because the equations of general relativity describe spacetime as a field and because those equations match experiment to a very high degree of accuracy, their interpretation of those equations *must* be true. But if quantum mechanics has taught us anything, it is to be wary of interpretation. The fact is, in quantum mechanics, we have equations that match experiment to a very high degree of accuracy but we have many interpretations of those equations. Why is it that we assume our interpretation of relativity is correct? Clearly the mathematics is correct and because the symbols in the mathematics have meaning, there is an aspect of our interpretation that *must* be correct to an extent. But consider the fact that we debate the ontological status of the wavefunction but we *don't* debate the ontological status of gravitational waves. In both cases we have similar mathematics that matches similar experimental results (thanks to the recent successful observation of gravitational waves by LIGO [1]). Yet the recent LIGO results are likely to be taken by some physicists as further evidence that spacetime is a real, physical manifold rather than interpreting that manifold as a mathematical convenience. The fact is, general relativity has nothing to say about the ontological status of the manifold. It is entirely possible to interpret the equations of general relativity as simply defining the relations between gravitationally interacting bodies. After all, if gravitons exist and gravity fits in a minimally extended Standard Model as we expect it should, how do we reconcile this with the classical notion that gravity is merely a curvature in spacetime?

On the other hand, it seems to me that pure operationalist interpretations of physics miss the deeper unifying beauty and simplification that can accompany a given interpretation. While an operationalist might view Newton's laws and the laws of conservation of energy and momentum as being on equal footing, this misses the fact that the former are direct consequences of the latter. This kind of thinking relegates any search for a deeper truth to a minor role and seems to miss the point of physics in the first place. The universe isn't a collection of disconnected parts. Everything is interwoven in a rich tapestry of connected phenomena. Physics needs to acknowledge that. But it needs to do that in a scientifically honest way by acknowledging its limits and not over-selling its results.

To briefly summarize this section, I have noted that physics has undeniable limits that are fundamental and that will never be overcome. In many cases we already know what these limits are. Yet physics (and science in general) sometimes oversells and sensationalizes its results. It is imperative that we remain true to the methods that have proven so successful for so long. They have proven so durable precisely because they are fruitful in what they have allowed us to deduce about the world and because they have led to discoveries that have had a tremendous impact on humanity and the rest of the planet (sometimes for good and sometimes not). For them to remain durable they cannot be oversold and we, as scientists, must remain humble about our results but confident in our methodology.

## 1.6 Conclusion

In this essay I have endeavored to address some of the more pernicious misconceptions about science, and physics in particular, that continue to haunt us. I have laid out a hopefully compelling argument that science *does* uncover an element of 'truth' about the world, i.e. that there *is* an underlying objective reality to it. Indeed, such a reality *must* exist otherwise there is no point in any of this. It is simply absurd to think that the rules governing the foundation of the world we inhabit are literally changing under our feet, or can be arbitrarily chosen by different observers. The mere fact that we can communicate with one another is evidence against this. Denying all objectivity is a fantasy that discounts the tremendous advances we have made as a species even if some of those advances have not always been for good.

That said, to some extent science is, to slightly paraphrase Eddington, the rational correlation of our shared experience. Indeed, this objective reality that science uncovers exists in the relational aspect of physical laws. As such the 'structure' of reality lies in a method of its description. That structure finds its most elegant and concise description in the language of mathematics which is precisely that: a language. It arises from the need to find simple representations for and general similarities in the phenomena that we observe and measure. It is entirely self-consistent (though Gödel showed us that we can't actually fully prove that fact). And yet it is the wonder of mathematics that sometimes leads us to make claims that cannot be verified by experiment or observation.

Indeed, as I have also argued, science, has limits that must be acknowledged. These are fundamental limits that will not be overcome with a new theory or a few clever experiments. They simply can't be. While we may find that these limits are not as restrictive as we might have originally thought, we will never find that they are simply wrong, per se, without there being something very seriously wrong with both physics and mathematics. Given the preponderance of evidence over the span of close to four centuries that indicates that the methodology of science is correct, not to mention the several millennia of evidence in support of the self-consistency of mathematics and logic (Gödel's theorems not-withstanding), it is highly unlikely that a flaw exists at the fundamental level of either. That said, there can be flaws in our *interpretation* of physics and mathematics and so we must remain cautious in how we extrapolate meaning from our results.

The evidence of the success of science as a methodology is all around us, most often in the technological advances that either directly resulted from scientific discoveries or were only properly understood through science. We find it in the explosion in human life expectancy since the 19th Century, in the amazingly reliable predictions of most modern sciences, particularly the physical sciences, and in the fact that modern science has endured for four centuries. And that is precisely why it is imperative that we not lose sight of its boundaries. Acknowledging those boundaries requires balancing the objective truths that science does tell us with the fact that it also sometimes defines its own limits.

Science is quite possibly the greatest achievement in all of human history. It deserves a vigorous defense against attacks and mis-understandings both from within and without. We owe it to ourselves and to each other.

**Acknowledgements** I wish to thank Kevin Staley, Joe Troisi, David Banach, and Tom Moore for fruitful discussions about many of the points addressed in this essay. I also wish to acknowledge the Saint Anselm College Philosophy Club for inspiration, commentary, and good food.

# References

1. Abbott, B.P., et al.: Observation of gravitational waves from a binary black hole merger. Phys. Rev. Lett. **116**, 6 (2016)
2. Agnes, M.E. (ed.): Webster's New World College Dictionary. Wiley, Hoboken, NJ (2003)
3. Albert, D.Z., Barrett, J.A.: On what it takes to be a world. Topoi **14**, 35–37 (1995)
4. Bergmann, P.: The Riddle of Gravitation. Dover Publications, Inc., 1st reprint of 1968 edition (1992)
5. Boixo, S., Flammia, S.T., Caves, C.M., Geremia, J.M.: Generalized limits for single-parameter quantum estimation. Phys. Rev. Lett. **98** (2007)
6. Bohm, D.: Quantum Theory. Prentice Hall, New York (1951)
7. de Bruijn, N.G.: The mathematical vernacular, a language for mathematics with typed sets. Stud. Logic Found. Math. **133**, 865–935 (1994)
8. Brukner, C., Zeilinger, A.: Quantum physics as a science of information. In: Elitzur, A.C., Dolev, S., Kolenda, N. (eds.) Quo Vadis Quantum Mechanics?, pp. 47–61. Springer (2005)
9. Byrne, P.: Personal Communication (2016)
10. Cantlon, J.F.: Math, monkeys, and the developing brain. Proc. Nat. Acad. Sci. **109**, 10725–10732 (2012)
11. Clauser, J.F., Shimony, A.: Bell's Theorem: experimental tests and implications. Rep. Prog. Phys. **41**, 1881–1927 (1978)
12. D'Ariano, M.G., Manessi, F., Perinotti, P.: Determinism without causality. Physica Scripta **2014**, T163 (2014)
13. Dai, N.: Lecture 1: the history of mechanics. In: Lu, Y. (ed.) A History of Chinese Science and Technology, pp. 294–295. Springer/Shanghai Jiao Tong University Press, Heidelberg/Shanghai (2015)

14. Davies, P.C.W.: Why is the physical world so comprehensible? In: Zurek, W.H. (ed.) Complexity, Entropy and the Physics of Information, pp. 61–70. Addison Wesley, Redwood City (1990)
15. Deutsch, D.: Quantum Theory, the Church-Turing Principle and the Universal Quantum Computer. Proc. R. Soc. London A **400** (1985)
16. Durham, I.T.: Unification and Emergence in Physics: the Problem of Articulation. http://fqxi.org/community/essay/winners/2009.1#Durham (2009)
17. Durham, I.T.: In Search of Continuity: Thoughts of an Epistemic Empiricist. http://arxiv.org/abs/1106.1124 (2011)
18. Durham, I.T.: An order-theoretic quantification of contextuality. Information **5**, 508–525 (2014)
19. Eddington, A.S.: Relativity Theory of Protons and Electrons. Cambridge University Press, Cambridge (1936)
20. Eddington, A.S.: The Philosophy of Physical Science. Cambridge University Press, Cambridge (1939)
21. Eddington, A.S.: Fundamental Theory. Cambridge University Press, Cambridge (1946)
22. Elitzur, A.C., Vaidman, L.: Quantum mechanical interaction-free measurements. Found. Phys. **23**(7), 987–997 (1993)
23. Einstein, A.: Über das Relativitätsprinzip und die aus demselben gezogenen Folgerungen [On the relativity principle and the conclusions drawn from it]. Jahrbuch für Radioaktivität und Electronik **4**, 411–462 (1907)
24. Einstein, A.: Über den Einfluß der Schwerkraft auf die Ausbreitung des Lichtes. Annalen der Physik **35**, 898–906 (1911)
25. Einstein, A.: Lichtgeschwindigkeit und Statik des Gravitationsfeldes. Annalen der Physik **38**, 355–369 (1912)
26. Feynman, R.: The Character of Physical Law. British Broadcasting Corporation (1965)
27. Frege, G.: Begriffsschrift, eine der arithmetischen nachgebildete Formelsprache des reinen Denkens. Verlag von Louis Nebert, Halle (1879)
28. Ganesalingam, M.: The Language of Mathematics: A Linguistic and Philosophical Investigation. Springer, Heidelberg (2013)
29. Geroch, R., Hartle, J.B.: Computability and physical theories. Found. Phys. **16**(6), 533–550 (1986)
30. Giovannetti, V., Lloyd, S., Maccone, L.: Advances in quantum metrology. Nat. Photonics **5** (2011)
31. Ghirardi, G.C., Rimini, A., Weber, T.: A general argument against superluminal transmission through the quantum mechanical measurement process. Lettere al Nuovo Cimento **27**, 293–298 (1980)
32. Hall, M.J.W., Deckert, D.A., Wiseman, H.: Quantum phenomena modelled by interactions between many classical worlds. Phys. Rev. X **4** (2014)
33. Hamma, A., Markopoulou, F., Lloyd, S., Caravelli, F., Severini, S. Markstrom, K.: Quantum Bose-Hubbard model with an evolving graph as a toy model for emergent spacetime. Phys. Rev. D **81** (2010)
34. Jaffe, R.L., Jenkins, A., Kimchi, I.: Quark masses: an environmental impact statement. Phys. Rev. D **79** (2009)
35. Kuhn, T.S.: Objectivity, value judgment, and theory choice. In: The Essential Tension: Selected Studies in the Scientific Tradition and Change. University of Chicago Press (1977)
36. Kuhn, T.S.: The Structure of Scientific Revolutions. University of Chicago Press (1996)
37. Lewis, P.G., Jennings, D., Barrett, J., Rudolph, T.: The Quantum State Can be Interpreted Statistically. arXiv:1201.6554v1 (2012)
38. Magueijo, J.: New varying speed of light theories. Rep. Prog. Phys. **66** (2003)
39. Mann, C.R., Twiss, G.R.: Physics. Foresman and Co., Chicago (1910)
40. Mersini-Houghton, L.: Wavefunction of the universe on the landscape. **86**, 973–980 (2006)
41. Moore, T.A.: Six Ideas That Shaped Physics, Unit Q: Particles Behave Like Waves. McGraw-Hill, New York (2003)

42. Napolitano, M., Koschorreck, M., Dubost, B., Behbood, N., Sewell, R.J., Mitchell, M.W.: Interaction-based quantum metrology showing scaling beyond the Heisenberg limit. Nature **471**, 486–489 (2011)
43. Novaes, C.D.: Formal Languages in Logic: A Philosophical and Cognitive Analysis. Cambridge University Press, Cambridge (2012)
44. Pavlov, I.P.: Conditioned Reflexes. Dover Publications Inc, Mineola (2003)
45. Popescu, S., Rohrlich, D.: Causality and nonlocality as axioms for quantum mechanics. In: Hunter, G., Jeffers, S., Vigier, J.P. (eds.) Causality and Nonlocality as Axioms for Quantum Mechanics, pp. 383–389. Springer, Netherlands (1998)
46. Pusey, M.F., Barrett, J., Rudolph, T.: On the reality of the quantum state. Nat. Phys. advance online publication, http://www.nature.com/nphys/journal/vaop/ncurrent/abs/nphys2309.html#supplementary-information (2012)
47. Roy, S.M., Braunstein, S.L.: Exponentially enhanced quantum metrology. Phys. Rev. Lett. **100** (2008)
48. Sakurai, J.J.: Modern Quantum Mechanics. Addison Wesley Longman, Reading, Massachusetts (1994)
49. Scarf, D., Hayne, H., Colombo, M.: Pigeons on par with primates in numerical competence. Science **334**, 6063 (2011)
50. Schutz, B.F.: A First Course in General Relativity. Cambridge University Press, Cambridge (1990)
51. Seiberg, N.: Emergent spacetime. In: 23rd Solvay Conference in Physics (2005)
52. Shimony, A.: Events and processes in the quantum world. In: Penrose, R., Isham, C.J. (eds.) Quantum Concepts in Space and Time, pp. 182–203. Oxford University Press, Oxford (1986)
53. Tegmark, M.: The mathematical universe. Found. Phys. **38**(2), 101–150 (2008)
54. Turing, A.M.: On computable numbers, with an application to the entscheidungsproblem. Proc. London Math. Soc. **1**(230–265), s2–42 (1937)
55. Wallace, D.: Worlds in the Everett Interpretation. Stud. Hist. Philos. Mod. Phys. **33**, 637–661 (2002)
56. Weinberg, S.: Gravitation and Cosmology. Wiley (1972)
57. Weinberg, S.: To Explain the World: The Discovery of Modern Science. HarperCollins Publishers, New York (2015)
58. Weinberg, S.: Reflections of a whig physicist. American Physical Society March Meeting (2016)
59. Weinfurtner, S.: Emergent spacetimes. Ph.D. Thesis, Victoria University, Wellington, New Zealand (2007)
60. Wheeler, J.A.: World as System Self-Synthesized by Quantum Networking
61. Wheeler, J.: Information, physics, quantum: the search for links. In: Zurek, W.H. (ed.) Complexity, Entropy and the Physics of Information, pp. 3–28. Addison Wesley, Redwood City (1990)
62. Will, C.M.: Theory and Experiment in Gravitational Physics. Cambridge University Press, Cambridge (1995)
63. Zee, A.: Quantum Field Theory in a Nutshell. Princeton University Press, Princeton (2003)
64. Zwierz, M., Pérez-Delgado, C.A., Kok, P.: General optimality of the Heisenberg limit for quantum metrology. Phys. Rev. Lett. **105** (2010)
65. Zwierz, M., Pérez-Delgado, C.A., Kok, P.: Ultimate limits to quantum metrology and the meaning of the Heisenberg limit. Phys. Rev. A **85** (2012)

# Chapter 2
# Eddington's Limits of Knowledge: The Role of Religion

**Matthew Stanley**

## 2.1 Thinking About Science and Religion

When thinking about the limits of scientific knowledge, religion is often invoked. Religious ideas are outside science, perhaps; or religious dogma constrains what science can know. We must remember, though, that science and religion are not abstract categories. Science is done by scientists, and religions only exist in the practices and beliefs of their membership. So, one way to think about science and religion is through the individuals involved. Many scientists are, and have been, religious. Does that matter? Is there any role for religion when we are talking about science?

This is an empirical question, and we need to look for evidence to understand what is happening. A.S. Eddington, as a religious scientist deeply concerned with the limits of knowledge, may be a helpful case study. He engaged with a range of scientific disciplines, was thoughtful about the philosophical implications of his work, and wrote extensively on science and religion. He was also active at a time of profound shifts in the role of religion in the western world, and he can help us understand some of those transitions.

One of the advantages of looking closely at an individual is that we are immediately pulled away from very broad terms such as 'science' and 'religion.' Eddington was not just 'religious,' he was a Quaker at a very particular time and place. We cannot assume 'religion' means something general like belief in God. For example, if we are concerned with sources of religious truth, we must distinguish between divine inspiration, scriptural literalism, and the apostolic succession. These are all sources of religious knowledge, but adherents of different traditions (or even the same traditions in different eras) might disagree profoundly about their significance.

M. Stanley (✉)
New York University, New York, USA
e-mail: ms5100@nyu.edu

© Springer International Publishing Switzerland 2017
I.T. Durham and D. Rickles (eds.), *Information and Interaction*,
The Frontiers Collection, DOI 10.1007/978-3-319-43760-6_2

We need to be specific about what beliefs and practices we are interested in. For Eddington's case, we need to understand something about late 19th-century/early 20th-century Quakerism. The Quakers are a small Protestant sect dating back to the 17th century that emphasizes the presence of God within everyone, and a related embrace of mysticism, pacifism, and social activism. They are distinctive for their rejection of clergy, rituals, and many of the outward trappings of organized religion. Even scripture, while valued, takes second place to personal religious experience. Eddington himself was a product of a particular historical manifestation of the Quakers: the so-called 'Quaker renaissance' of the late Victorian period, which was characteristic of liberal theological movements of the time.

This movement was a result of a new generation of Quakers debating how to react to the march of modernism: science, history, industrialization, urbanization, pluralism. They decided to embrace modernism, with all of its flaws, but to try to invest those modern views with religious values. The question of how to interface with science was particularly fraught. One Quaker leader wrote:

> This theory of the detachment of science and religion from one another never has been a working theory of the universe; the two areas must overlap and blend, or we are lost [5, p. 219].

Eddington learned how to be a Quaker from exactly this group. He was taught from a young age that a scientist could not separate his religious beliefs and practices from his work. He became what one influential Quaker called a 'practical mystic': someone who lived and worked in the temporal world but who was still in direct experiential contact with spiritual forces.

## 2.2 Types of Interactions

Once we have established the specific religious categories we are interested in, we can start asking more detailed questions. In what ways might we imagine these religious categories interacting with scientific work? We are focused here primarily on conceptual issues, and this discussion will address four broad categories (with some overlap) that can help shape scientific concepts and methods: restriction, inspiration, natural theology, and values.

### 2.2.1 Restriction

We often think about religious thought functioning as a bias against certain scientific concepts. This was not particularly an issue for Eddington, but Quakers of earlier generations were hostile to Darwin and notions of human evolution. Human evolution is perhaps the classic example of an idea resisted for religious reasons. Even beyond caricatures such as Creation Scientists, well-respected productive scientists such as William Thomson (Lord Kelvin) found themselves unwilling to accept natural selection [11].

However, this sort of restriction sometimes happens in more subtle ways. Religious beliefs often involve ways of thinking and acting, not just statements of fact about the material world. Michael Faraday was famously skeptical of hypothetical reasoning and speculative theory, a position likely linked to his idiosyncratic theological positions regarding humility before God [1].

## 2.2.2 Inspiration

If we accept that religion can shape scientific thinking in negative ways, it should come as no surprise that the opposite can happen as well. A theological viewpoint may provide a scientifically useful idea, or make a particular theoretical perspective more appealing. To take our earlier example, Kelvin's religious beliefs were a critical part of how he engaged with the second law of thermodynamics. He felt that there was strong scriptural support for a universe which was gradually running down, and saw entropy as an affirmation of a divinely created universe. Similarly, Maxwell was convinced that the Christian God had designed the laws of nature with unity and simplicity in mind. This was a major stimulus for his research program that would eventually unify electricity, magnetism, and light [6, 10].

In Eddington's case, we can see this at work in his embrace of the implications of quantum physics. As a Quaker, he believed strongly in the reality of free will and the ability of humans to make meaningful choices in the world. He had already rejected determinism, so he was particularly receptive to quantum indeterminism. It is important to note that this was not a matter of him thinking that quantum physics had proved his religious beliefs to be true. Rather, his religious beliefs provided a receptive environment in which the oddities of the quantum world were expected and welcome.

## 2.2.3 Natural Theology

A related, though distinct, process is referred to as natural theology. This describes practices and beliefs involved with the idea that scientific study can reveal religious truths. It is sometimes described as the assertion that there is a book of nature just as there is a book of scripture, and both should be studied to learn about God. This was a common motivation for scientific practice before the twentieth century, particularly in Britain. Natural theology was often concerned with questions of proof—can science provide evidence for God's existence, or the truth of the Bible? The hope (usually unfulfilled) was that science could so surely demonstrate divine creation and foresight that every atheist would be forced to capitulate.

Eddington, like most Quakers of his generation, rejected the reasoning of natural theology. For them, proof was not an important issue in religion. Religion was an ongoing, always developing realm of human experience, and should not be expected to provide certainty. When Eddington became well known as a religious scientist, he bemoaned being pressured to use science to prove the truth of Christianity. He complained that believers were constantly asking him to provide a scientific silver bullet to use against skeptics. In response, he disclaimed that he could no more force belief into an atheist than he could ram a joke into a Scotsman [4, p. 336]. Proof was simply not a part of his theology.

Quakers talked to God not through nature, but through direct, individual, internal experience. Eddington tried to illustrate this through a quote from scripture. In the first book of Kings, God came to communicate with Elijah:

> And he said, Go forth, and stand upon the mount before the LORD. And, behold, the LORD passed by, and a great and strong wind rent the mountains, and brake in pieces the rocks before the LORD; but the LORD was not in the wind: and after the wind an earthquake; but the LORD was not in the earthquake: And after the earthquake a fire; but the LORD was not in the fire: and after the fire a still small voice And, behold, there came a voice unto him, and said, What doest thou here, Elijah? [2, pp. 25–26].

God was not to be found in meteorology, seismology, or physics, but inside oneself, through mystical experience. Eddington was deeply religious but rejected the claim of natural theology that God should be invoked in science.

## 2.3  Science and Values

The previous types of interactions became less prevalent and influential in the practice of science in the twentieth century. All of them are largely concerned with the truth claims of religion—did God create the world? In what way? Are humans separate from the natural order? However, both religion and science are more than a collection of truth claims. Both are also collections of values—attitudes, preferences, ways of thinking or acting. Tolerance is a value. Precision is a value. Some values appear in both religious and scientific contexts, and can provide a way for both traditions to interact in interesting ways. In analogy to chemical bonding, I call these *valence values* [9, pp. 5–7]. The fact that they are shared can bring together otherwise disparate practices. Valence values are the best way to understand Eddington's work as a religious scientist, particularly his thinking on the limits of science. In this section I will examine two groups of values that, for Eddington, were important in both science and religion: pacifism, and open-minded seeking.

## 2.3.1 Pacifism and Internationalism

The Quakers are perhaps best known for their pacifism and commitment to internationalism. That value is rooted in a religious principle called the peace testimony. Their 'belief of the potentiality of the divine in all men' led them to reject war and violence in all its forms. Attacking another person was, essentially, attacking God. The Quaker Renaissance saw an important shift in their pacifism. Whereas previous generations simply refused to fight, Eddington's generation felt called to be activists for peace—working constantly to prevent conflict and repair rifts between nations.

This was particularly important for Eddington due to the way the scientific community reacted during World War I. After the outbreak of war, British scientists rejected anything that looked German. Foreign members were thrown out of the Royal Society, astronomical telegraph lines were cut, scientific societies refused to send journals to enemy countries. Eddington was one of the very few British scientists who resisted these moves. He protested jingoism in science both on practical grounds (the lines of latitude and longitude do not obey national boundaries) and higher ideals (the pursuit of truth is sacred beyond any patriotic feuds).

Unlike most of his colleagues, Eddington maintained communications with scientists in enemy and neutral countries. This meant that he was the only person in Britain to hear about Einstein's new theory of general relativity in 1915. Eddington was excited by the theory's scientific and philosophical significance, and even further when he learned that Einstein was himself a pacifist. Promoting Einstein and his ideas could promote world peace and refute the racist stereotypes that were fueling wartime hatred.

This helps us understand the intensity of Eddington's efforts to carry out the 1919 eclipse to test Einstein's prediction of light deflection. It required exactly the international cooperation that Eddington had argued was fundamental to the spirit of science: German theory, British observers, Brazilian and Portuguese sites. Further, it was an opportunity to bring a peace-loving, insightful German to prominence in both science and society and thus weaken the prejudice that had shattered international science. To Eddington the expedition was not only part of his duty as a scientist (it would test a theory of radical physical implications), but he felt a linked duty as a Quaker (it would help repair international relations by bringing together the German and British scientific communities) [7, 8].

Eddington used the spectacular publicity surrounding the expedition to lay out his case for repairing international relations both inside and outside science. To do so, he adopted the techniques used by Quaker relief workers during and after the war. For him, Einstein and relativity were his contribution as a Quaker to world peace. The key strategy was to build intellectual relationships, and to humanize the enemy—to show that they were just like us, and suffered just as much from war. Eddington's religious values were key to this scientific project, and drew stark contrast with his British colleagues who did not share his pacifist perspective.

## 2.3.2 Seeking

Eddington's scientific work was also shaped by a Quaker attitude toward knowledge called 'seeking'. This is an outlook on religious experience that emphasizes constant exploration and searching for new things. Seeking is essentially an anti-dogmatic position: do not look for complete certainty, because that leads to stagnation. This is derived from the Quakers' mystical practices. If you talk to God every day, you need to always be receptive to new knowledge. In contrast, a fundamentalist who emphasizes unchanging scripture comes to value stability and certainty over novelty. Quaker seeking embraces progress despite uncertainty, and resulted in a kind of mystical pragmatism. Whatever tools were useful for spiritual progress were to be celebrated, even if their ultimate meaning was unclear.

Eddington argued that this sort of attitude was not just useful in religion, but also in science. Particularly in his work on stellar physics we can see Eddington constantly emphasizing this need to embrace incomplete knowledge and to reject complete certainty as a goal. He began work on stellar astrophysics before many of the fundamental nuclear processes were understood, and investigators had been continually frustrated in their efforts to model the interiors of stars. Many physicists, such as James Jeans, argued that if they did not know every principle at work there, and every detail of the star's history, they could know nothing at all about what was happening. Eddington instead shaped his work around the idea that we can use incomplete, uncertain theories to solve specific problems. He became famous at the Royal Astronomical Society for his shrewd use of assumptions, approximations, and outright guesses to evade difficult theoretical problems and arrive at some useful conclusion. This is the sort of methodology that led to his development of the mass-luminosity relation, one of the first real inroads to understanding stellar structure. This eventually provided traction on the problem of stellar energy, even as the full understanding of fusion was over a decade away. Some, like Jeans, complained that this partial knowledge should not really count as science. In contrast, Eddington tried to convince colleagues that exploring a new problem, even incompletely, was valuable unto itself. His case was that tentative knowledge was welcome as long as it led to progress [9].

This represented an extension of Eddington's Quaker attitude toward religious knowledge, in which fundamental certainty (such as inerrant scripture) was far less important than maintaining a living, transforming faith and a direct experience of God. Certainty was not to be sought after in either science or religion. In one of his best-selling books, Eddington wrote:

> In science as in religion the truth shines ahead as a beacon showing us the path; we do not ask to attain it; it is better far that we be permitted to seek You will understand neither science nor religion unless seeking is placed in the forefront [2, pp. 22–23].

He said science and religion were similar in that they were both an unending quest. Further, this urge to seek was the deepest expression of human nature—science came from as profound a root as religion.

## 2.4 Experience

The most important of Eddington's valence values for our discussions here was that of *experience*. Quakers had a distinctive way of thinking about how humans experience the world, and this informed Eddington's approach to relativity and quantum physics. In his extensive writings on relativity he always emphasized the role of the observer. His epistemological emphasis eventually developed into his 'selective subjectivism.' This essay will look at the roots and implications of this framework as opposed to the framework itself.

In interpreting relativity, Eddington particularly stressed the observer's role in making measurements. He even elevated measurability (or he sometimes wrote, 'metricality') to be the very marker of scientific knowledge. The essence of relativity was, he said, simply a rigorous treatment of this idea. What are we measuring, and how do we measure it?

Pushing this even further, Eddington wanted to distill these observations to simple relations. He drew extensively on his intellectual genealogy. His reading of Karl Pearson helped shape his emphasis on the role of conscious minds in organizing distinct sense impressions. He explicitly credited Bertrand Russell as his most important philosophical influence, and it is not difficult to see that philosopher's structuralism in Eddington's work.

Eddington's definition of science was essentially Einstein as filtered through Pearson and Russell. Exact science was simply the symbolic analysis of pointer readings collected during space-time events (that is, intersections of world-lines). Thus science was limited to a structural relationship of symbols and binary events. Relativity and quantum mechanics were the pinnacle of modern science because they embraced this operationalism and discarded deterministic materialism. Even further, both theories acknowledged the importance of human consciousness.

Eddington contended that this move, the rigorous analysis of the act of observation, was the foundation of scientific inquiry. He stressed that observation was the human mind selecting data from the four-dimensional universe (what Eddington called 'the World'). And the selection process was not passive. Rather, our minds chose which elements of reality were important and worth analyzing. The World was a vast soup of space-time events which could be sliced up and perceived in many different ways. The observer's selection of a particular slice results in their perception of certain laws, such as conservation of mass. Our minds like permanence, so they select for conservation laws that create that effect. Einstein's law of gravity came directly from these selection effects. The laws of physics were therefore created by human observation. This 'closed cycle' of physics, in which the mind both creates and discovers natural laws, gave humanity and human experience a renewed place of importance. Eddington stressed that this demonstrated that our world was fundamentally idealistic, and the rigid materialism beloved of skeptics was no longer tenable.

This left humanity with two worlds of experience: one metrical, one non-metrical. The world of science was completely metrical. It could only speak about the measurable aspects of reality. But there were aspects of human experience that were not quantifiable (love, beauty). Relativity's emphasis on measurements and coincidences acknowledged a world beyond that of measurable science. Eddington illustrated this with his famous tale of two tables [4]. He described himself sitting and writing at two different tables. One was the table of science. It was mostly emptiness, made of atoms whirling to and fro, without any substance or stability. The other was the table of ordinary experience. It was solid and steady. Science could only describe the former; a person could only write on the latter. The solid table came from the direct experience of our consciousness. The scientific table was filtered through pointer readings and abstract symbols. The scientific table was, therefore, far less *real* than the other. This is the root of 'Eddington's challenge' as discussed by Huw Price at the conference that inspired this volume. We have a sense of time provided by our ordinary experience, and a sense of time provided by scientific analysis. Are they compatible? Should one be valued over the other?

Eddington contended that this sort of ordinary experience was really *spiritual* experience. This spiritual world was the one in which everyone actually lived. It was the world marked by our intuitive convictions of experience and consciousness. Spiritual reality and scientific reality were separated by the border of metricality. Relativity's emphasis on measurement thus acknowledged the reality of spiritual experience.

This was, Eddington thought, good for both science and religion. Emphasizing the role of direct experience affirmed the spiritual life and created modern physics. He wrote that this spiritual outlook would benefit science:

> The anti-materialistic attitude of religion would certainly be an advantage to modern science. It would help a man to see certain possibilities, to entertain certain speculations, which the old materialists, who regarded the universe as composed of little billiard balls, would have difficulty in grasping [2, pp. 50–51].

This emphasis on the spiritual value of experience was particularly Quaker. Eddington's attack on materialistic critiques of religion was typical of British liberal theology. It was notably Quaker, though, in emphasizing individual, non-dogmatic experience. For Eddington, both the metrical and spiritual worlds were empirical:

> The scientist and the religious teacher may well be content to agree that the value of any hypothesis extends just so far as it is verified by actual experience [4, p. 222].

The scientific and spiritual worlds were quite different, but they both emerged from the same values.

It is important to note that this was not a replacement religion—Eddington did not want anyone to worship covariance. Neither did he think that relativity confirmed Quakerism. Rather, his argument was that this interpretation simply opened up a space in which an individual's own religious experience could be accepted on its own terms without scientific critique.

Despite this, it is common to see Eddington quoted as saying 'religion first became possible in 1927' (referring to the development of Heisenberg's uncertainty principle). The quote is usually presented to support the claim that quantum physics proved the validity of religion. Amazingly, this is exactly the opposite of the meaning of Eddington's actual statement. The original quote reads as follows:

> It will perhaps be said that the conclusion to be drawn from these arguments from modern science, is that religion first became possible for a reasonable scientific man about the year 1927. If we must consider that most tiresome person, the consistently reasonable man, we may point out that not merely religion but most of the ordinary aspects of life became possible for him in that year [4, p. 350].

He was very clear that 1927 made nothing new possible for religion. Instead, modern physics was just providing reassurance to religious people that their experiences of the spiritual world could be valid on their own terms. Eddington expected everyone to go on having religious experiences just as they always had been. Relativity provided the critical first step by recognizing the importance of mind and consciousness, but any further religious guidance had to come from direct individual spiritual experience. What modern physics did was make a particular kind of religion more plausible—Quakerly experiential religion. Scripturalists and High Church advocates were given little help.

## 2.5 Information and Interaction

Two of the themes of this volume are information and interaction, each critical concepts for understanding Eddington's limitations of science. The role of "interaction" is very clear—Eddington's entire epistemology of science is built on intersections of world-lines and the interaction of the mind with the World. Considering the role of 'information' is both more hazy and perhaps more rewarding. Eddington certainly uses metaphors and terms that evoke modern senses of information theory: observations are 'sets of signals passing along our nerves,' and pointer readings are 'code messages' that have no meaning without a conscious observer [3, p. 6]. Deciphering the messages of the outside world is a metaphor he returns to again and again. The scientific observer is one who tries to decode the patterns seen in those messages.

In an important sense, Eddington denied that there is any scientific reality to the physical world other than what we can discern from these signals. He reduces those signals to their most primitive forms—pointers read by a one-eyed observer with no color vision. He would not have used the term, but we can think of his observations as bits. To Eddington, this reduction of the entire scientific world to bits was the foundation of the geometrization of physics begun by Einstein. In this sense, Eddington might have said that information is distinctive of physical science, but not experience generally. The non-metrical world of spiritual experience could not be reduced to bits the same way. Regardless, understanding both categories was a matter of epistemology, which perhaps does put information at the root.

## 2.6   Conclusion

There are many ways of thinking about science and religion. This essay has examined some modes that are useful for understanding Eddington's approach to the limits of science. Even in situations such as this where the importance of religion is clear, it is rarely the sole explanation. Religious beliefs and practices necessarily interact with many other factors.

The case of Eddington shows that sometimes we do see real interactions between religion and science. This can make some ideas more plausible (quantum physics); some ideas less (determinism); some approaches better (seeking), some worse (nationalism). Sometimes there is no direct interaction even where we might expect it. For example, Eddington's religiosity did not seem to have any links to his cosmological work on the beginning of the universe. This should not be a surprise. Quakers of his generation were not particularly interested in questions of creation and Genesis. Finding that sometimes science and religion interact does not mean they always do.

Further, these interactions are historically situated. They appear in some times and places and not others. What we see with Eddington is distinctive of early twentieth century liberal theology. We would not expect to see the same valences in nineteenth century Britain, which was dominated by robust institutional religion. It is also different today. The strong Protestant fundamentalism in modern America has completely different values than Quakerism, and generally struggles with science. And it will be different in the future. We need to be careful about how we generalize. When talking about science and religion, we need to pay attention to the particulars.

## References

1. Cantor, G.: Michael Faraday: Sandemanian and Scientist. Macmillan, Basingstoke (1991)
2. Eddington, A.S.: Science and the Unseen World. Macmillan Company, New York (1929)
3. Eddington, A.S.: New Pathways in Science. Cambridge University Press, Cambridge (1935)
4. Eddington, A.S.: The domain of physical science. In: Needham, J. (ed.) Science, Religion and Reality, 2nd edn, pp. 193–224. George Braziller, New York (1955)
5. Report of the Proceedings of the Conference of Members of the Society of Friends. Headley Bros, London (1896)
6. Smith, C.: The Science of Energy. University of Chicago Press, Chicago (1998)
7. Stanley, M.: An expedition to heal the wounds of war: the 1919 eclipse expedition and eddington as quaker adventurer. Isis **94**, 57–89 (2003)
8. Stanley, M.: Practical Mystic: Religion, Science, and A.S. Eddington. University of Chicago Press, Chicago (2007)
9. Stanley, M.: So simple a thing as a star: jeans, eddington, and the growth of astrophysical phenomenology. Br. J. Hist. Sci. **40**, 53–82 (2007)
10. Stanley, M.: Huxley's Church and Maxwell's Demon. University of Chicago Press, Chicago (2015)
11. Wise, M.N., Smith, C.: Energy and Empire: A Biographical Study of Lord Kelvin. Cambridge University Press, Cambridge (1989)

# Chapter 3
# Eddington's Dream: A Failed Theory of Everything

**Helge Kragh**

## 3.1 Introduction

Arthur Stanley Eddington is recognized as one of the most important scientists of the first half of the twentieth century [9]. He owes this elevated position primarily to his pioneering work in astronomy and astrophysics, and secondarily to his expositions of and contributions to the general theory of relativity. Eddington developed a standard model of the interior structure of stars and was the first to suggest nuclear reactions as the basic source of stellar energy. In 1919 he rose to public fame when he, together with Frank Dyson, confirmed Einstein's prediction of the bending of starlight around the Sun. Six years later he applied general relativity to white dwarf stars, and in 1930 he developed one of the first relativistic models of the expanding universe known as the Lemaître-Eddington model.

Not satisfied with his accomplishments within the astronomical sciences, during the last part of his life Eddington concentrated on developing an ambitious theory of fundamental physics that unified quantum mechanics and cosmology. The present essay is concerned solely with this grand theory, which neither at the time nor later won acceptance. Although Eddington's fundamental theory has never been fully described in its historical contexts, there are several studies of it from a philosophical or a scientific point of view (e.g. [8, 25, 26]). There have also been several attempts to revive scientific interest in Eddington's theory, such as [37], but none of them have been even moderately successful. At any rate, this essay is limited to the theory's historical course.

H. Kragh (✉)
University of Copenhagen, Copenhagen, Denmark
e-mail: helge.kragh@nbi.ku.dk

© Springer International Publishing Switzerland 2017
I.T. Durham and D. Rickles (eds.), *Information and Interaction*,
The Frontiers Collection, DOI 10.1007/978-3-319-43760-6_3

## 3.2   Warming up

Although Eddington only embarked on his ambitious and lonely research programme of unifying the atom and the universe in 1929, some of the features of this programme or theory can be found in his earlier work. When Dirac's wave equation of the electron opened his eyes to a new foundation of physics, he was well prepared.

The significance of combinations of the constants of nature, a key feature in the new theory, entered Eddington's masterpiece of 1923, *The Mathematical Theory of Relativity*. Referring to the 'very large pure number [given] by the ratio of the radius of the electron to its gravitational mass $= 3 \cdot 10^{42}$,' Eddington [12, p. 167] suggested a connection to 'the number of particles in the world—a number presumably decided by pure accident.' And in a footnote: 'The square of $3 \cdot 10^{42}$ might well be of the same order as the total number of positive and negative electrons' (see also [10, p. 178]). Of course, Eddington's reference to the positive electron was not to the positron but to the much heavier proton, a name introduced by Rutherford in 1920 but not generally used in the early 1920s. While Eddington in 1923 thought that the number of particles in the world was accidental, in his later theory this 'cosmical number' appeared as a quantity determined by theory.

A somewhat similar suggestion had been ventured by Hermann Weyl a few years earlier when considering the dimensionless ratio of the electromagnetic ratio of the electron ($r_e = e^2/mc^2$) to its gravitational radius ($r_g = Gm/c^2$). According to Weyl [38], the ratio of the radius of the universe to the electron radius was of the same order as $r_e/r_g \approx 10^{40}$, namely

$$\frac{r_e}{r_g} = \frac{e^2}{Gm^2} \approx \frac{R_E}{r_e} \tag{3.1}$$

where $R_E$ denotes the radius of the closed Einstein universe. The huge value of the electrical force relative to the gravitational force as given by $e^2/Gm^2$ had been pointed out much earlier (e.g. [3]), but it was only with Weyl and Eddington that the number ca. $10^{40}$ was connected to cosmological quantities.

Another feature of Eddington's mature philosophy of physics, apart from its emphasis on the constants of nature, was the fundamental significance he ascribed to mind and consciousness. Physics, he argued, can never reveal the true nature of things but only deals with relations between observables that are subjectively selected by the human mind. As Eddington [11, p. 155] stated in his earliest philosophical essay, he was 'inclined to attribute the whole responsibility for the laws of mechanics and gravitation to the mind, and deny the external world any share in them.' By contrast, 'the laws which we have hitherto been unable to fit into a rational scheme are the true natural laws inherent in the external world, and mind has no chance of moulding them in accordance with its own outlook.' The same theme appeared prominently in his Gifford Lectures delivered in early 1927, where Eddington [13, p. 281] provocatively concluded that 'the substratum of everything is of mental character.'

While up to this time Eddington had exclusively relied on general relativity, he now also appealed to the new symbolic quantum theory as a further argument in favour of his view that physicists manufacture the phenomena and laws of nature. Yet, although he took notice of Heisenberg's and Schrödinger's quantum mechanics, he was uncertain of how to make sense of it within the framework of a unified theory of relativistic physics. In *The Mathematical Theory of Relativity* [12, p. 237] he wrote: 'We offer no explanation of the occurrence of electrons or of quanta. The excluded domain forms a large part of physics, but it is one in which all explanation has apparently been baffled hitherto.' Four years later the situation had not changed materially.

## 3.3  Eddington Meets the Dirac Equation

When Eddington studied Paul Dirac's relativistic wave equation of the electron in early 1928 he was fascinated but also perplexed because it was not written in the language of tensor calculus [19, p. 2]. Although he recognized the new quantum equation as a great step forward in unifying physics, he also thought that it did not go far enough and consequently decided to generalize it. For an electron (mass $m$, charge $e$) moving in a Coulomb field the positive-energy Dirac equation can be written

$$\frac{ih}{2\pi}\frac{\partial\psi}{\partial t} = \frac{e^2}{r}\psi + c\sqrt{(ih/2\pi)^2\Delta + m^2c^2}\psi \tag{3.2}$$

where $\Delta = (\partial^2/\partial x^2, \partial^2/\partial y^2, \partial^2/\partial z^2)$ is the Laplace operator. Eddington [14, 15] rewrote the equation by introducing two constants, namely

$$\alpha^{-1} = \frac{hc}{2\pi e^2} \tag{3.3}$$

$$\gamma = \frac{2mc}{h} \tag{3.4}$$

In this way he arrived at

$$-\alpha\frac{\partial\psi}{\partial t} = \frac{ic}{r}\psi + c\sqrt{\alpha^2\Delta - \gamma^2}\psi \tag{3.5}$$

The constant $\alpha^{-1}$ is the inverse of what is normally called the fine-structure constant, but Eddington always (and somewhat confusingly) reserved the name and symbol $\alpha$ for the quantity $hc/2\pi e^2$. In his initial paper of 1929 he applied his version of the Dirac equation and his unorthodox understanding of Pauli's exclusion principle to derive the value

$$\alpha^1 = 16 + \frac{1}{2} \times 16 \times (16 - 1) = 136 \tag{3.6}$$

By 1929 the fine-structure constant was far from new, but it was only with Eddington's work that the dimensionless combination of constants of nature was elevated from an empirical quantity appearing in spectroscopy to a truly fundamental constant [28]. Moreover, Eddington was the first to focus on its inverse value and to suggest—indeed to insist—that it must be a whole number. He was also the first to argue that $\alpha$ was of deep cosmological significance and that it should be derivable from fundamental theory.

When Eddington realized that the theoretical value $\alpha^{-1} = 136$ did not agree with experiment, at first he pretended to be undisturbed. 'I cannot persuade myself that the fault lies with the theory,' he wrote in his paper of 1929. All the same, as experiments consistently showed that $\alpha^{-1} \cong 137$ he was forced to look for a fault in his original theory. He soon came to the conclusion that $\alpha^{-1} = 137$ *exactly*, arguing that the extra unit was a consequence of the exclusion principle, which in his interpretation implied the indistinguishability of any pair of elementary particles in the universe. For the rest of his life Eddington [16, p. 41] stuck to the value 137, which he claimed to have 'obtained by pure deduction, employing only hypotheses already accepted as fundamental in wave mechanics.'

It should be pointed out that although Dirac's linear wave equation served as the inspiration for Eddington's grand theory, for him it was merely a temporary stepping stone towards a higher goal. He felt that relativistic quantum mechanics, whether in Dirac's version or some of the other established versions, was still characterized by semi-empirical methods that prevented a rational foundation of laws unifying the micro-cosmos and the macro-cosmos. In his monograph *Relativity Theory of Protons and Electrons* [19, pp. 6–7] Eddington emphasized the difference between his theory and the one of Dirac: 'Although the present theory owes much to Diracs theory of the electron, it differs fundamentally [from it] on most points which concern relativity. It is definitely opposed to what has commonly been called 'relativistic quantum theory,' which, I think, is largely based on a false conception of the principles of relativity theory' (Fig. 3.1).

**Fig. 3.1** Dirac (*third from left*) with Eddington and Schrödinger (*third and second from right*) at a colloquium in 1942 at the Dublin Institute for Advanced Studies. *Source* http://www-history.mcs. st-and.ac.uk/Biographies/Tinney.html

## 3.4 Constants of Nature

The fine-structure constant was not the only constant of nature that attracted the attention of Eddington. On the contrary, he was obsessed by the fundamental constants of nature, which he conceived as the building blocks of the universe and compared to the notes making up a musical scale: 'We may look on the universe as a symphony played on seven primitive constants as music played on the seven notes of a scale' [18, p. 227]. The recognition of the importance of constants of nature is of relatively recent origin, going back to the 1880s, and Eddington was instrumental in raising them to the significance they have in modern physics [1], [30, pp. 93–99]. Whereas the *fundamental* constants, such as the mass of an electron, are generally conceived to be irreducible and essentially contingent quantities, according to Eddington this was not the case. Not only did he believe that their numerical values could be calculated, he also believed that he had succeeded in actually calculating them from a purely theoretical basis. For example, for Newton's gravitational constant he deduced $G = 6.6665 \times 10^{-11} \, \mathrm{N \, m^2 \, kg^{-2}}$, in excellent agreement with the experimental value known at the time, which he stated as $(6.670 \pm 0.005) \times 10^{-11} \, \mathrm{N \, m^2 \, kg^{-2}}$ [23, p. 105].

The constants of nature highlighted by Eddington were the mass $m$ and charge $e$ of the electron, the mass of the proton $M$, Plancks constant $h$, the speed of light $c$, the gravitational constant $G$, and the cosmological constant $\Lambda$. To these he added the cosmical number $N^*$ and, on some occasions, the number of dimensions of space-time $(3 + 1)$. Two of the constants were original to him and deserve mention for this reason. The cosmological constant introduced by Einstein in his field equations of 1917 was not normally considered a constant of nature and was in any case ill-regarded by many cosmologists and astronomers in the 1930s. With the recognition of the expanding universe most specialists followed Einstein in declaring the cosmological constant a mistake, meaning that $\Lambda = 0$. Eddington emphatically disagreed. He was convinced that the $\Lambda$-constant was indispensable and of fundamental importance, for other reasons because he conceived it as a measure of the repulsive force causing the expansion of the universe. Appealing to Einstein's original relation between the constant and the radius of the static universe,

$$\Lambda = \frac{1}{R_E^2} \tag{3.7}$$

he considered $\Lambda$ to be the cosmic yardstick fixing a radius for spherical space. 'I would as soon think of reverting to Newtonian theory as of dropping the cosmical constant,' Eddington [17, p. 24] wrote. To drop the constant, would be 'knocking the bottom out of space.' Whereas the theoretical value of the cosmological constant is today one of physics' deep and unsolved problems, to Eddington it was not. In 1931 he provided an answer in terms of other constants of nature:

$$\Lambda = \left(\frac{2GM}{\pi}\right)^2 \left(\frac{mc}{e^2}\right)^4 = 9.8 \times 10^{-55}\,\text{cm}^{-2} \tag{3.8}$$

While this was believed to be of roughly the right order, unfortunately (or fortunately for Eddington) there were no astronomical determinations of $\Lambda$ with which the theoretical value could be compared. Eddington used the value of $\Lambda$ to calculate from first principles the Hubble recession constant, for which he obtained $H_0 = 528\,\text{km s}^{-1}\,\text{Mpc}^{-1}$. Since the figure agreed nicely with the generally accepted observational value, he took it as evidence of the soundness of his theoretical approach.

In his 1929 paper Eddington showed that the fine-structure constant was related in a simple way to constants of a cosmological nature, such as given by the expression

$$\alpha = \frac{2\pi mc R_E}{h\sqrt{N^*}} \tag{3.9}$$

He considered the cosmical number $N^*$—the number of electrons and protons in the closed universe—to be a most important constant of nature. According to conventional physics and cosmology there was nothing special about the number, which might well have been different, but Eddington not only insisted that it was a constant, he also claimed that it could not have been different from what it is. Moreover, he claimed that he was able to deduce $N^*$ rigorously from theory, just as he was able to deduce the other constants of nature. The result was

$$N^* = 2 \times 136 \times 2^{256} \cong 3.15 \times 10^{79} \tag{3.10}$$

Notice that Eddington gave the number precisely. He counted a positron as minus one electron, and a neutron as one proton and one electron. Assuming that the total number of electrons equals the number of protons, he further derived a relation between two very large dimensionless constants:

$$\frac{e^2}{GmM} = \frac{2}{\pi}\sqrt{N^*} \tag{3.11}$$

This was the relation that he had vaguely suggested as early as 1923. For the mass ratio $M/m$ between the proton and the electron, Eddington suggested that it could be found from the ratio of the two roots in the equation

$$10x^2 - 136\omega x + \omega^2 = 0 \tag{3.12}$$

The quantity $\omega$ is what Eddington called a 'standard mass,' the mass of an unspecified neutral particle. In this way he derived the theoretical value $M/m = 1847.6$ or nearly the same as the experimental value [8, pp. 211–218].

In the 1930s a few scientists speculated for the first time that some of the constants of nature might not be proper constants but instead quantities that vary slowly in time

[30, pp. 167–192]. Eddington considered such ideas to be pure nonsense. In 1938 Dirac proposed a cosmological theory based on the assumption that $G$ decreased according to

$$\frac{1}{G}\frac{dG}{dt} = -3H_0 \qquad (3.13)$$

However, Eddington [20] quickly dismissed Dirac's theory as 'unnecessarily complicated and fantastic.' He was not kinder to contemporary speculations that the speed of light might be a varying quantity: 'The speculation of various writers that the velocity of light has changed slowly in the long periods of cosmological time is nonsensical because a change in the velocity of light is self-contradictory' [23, p. 8]. Nearly sixty years later the so-called VSL (varying speed of light) theory revived the discussion of whether Eddington's objection was reasonable or not [24, 29].

## 3.5   Fundamental Theory

The research project that Eddington pursued with such fervour and persistence during the last 15 years of his life resulted in a long series of scientific papers and a couple of important monographs. Some of his books were highly technical while others were of a philosophical nature and mostly oriented toward a general readership. To Eddington, the latter were no less important than the first. He followed his research programme in splendid isolation, apparently uninterested in the work done by other physicists in the tradition he had initiated. The isolated and closed nature of Eddington's research is confirmed by bibliometric studies based on the list of publications given by his biographer, the Canadian astronomer Allie Vibert Douglas [9]. Among the references in Eddington's 14 research papers on his unified theory in the period 1929–1944, no less than 70% are to his own works. By comparison, the average self-reference ratio in physics papers in the period is about 10% [31].

The first major fruit of Eddington's efforts appeared in 1936 in the form of *Relativity Theory of Protons and Electrons* (RTPE) a highly mathematical and personal exposition of his ongoing attempt to create a new basis for cosmology and physics. During the following years he prepared a systematic account of his theory and its mathematical foundation, but *Fundamental Theory* only appeared after his death. The title was not Eddington's, but chosen but the mathematician Edmund Whittaker who edited Eddington's manuscript and supervised it to publication.

Whittaker had closely followed Eddington's work which fascinated him more from a philosophical than a physical point of view. Like most scientists he remained unconvinced about the physical soundness of the grand project. In an extensive review of RTPE, Whittaker [39] likened Eddington to a modern Descartes, suggesting that Eddington's theory did not describe nature any better than the vortex theory of the French rationalist philosopher. Nonetheless, he described Eddington as 'a man of genius.' Whittaker was not alone in comparing Eddington to Descartes. According to the philosopher Charlie Broad [2, p. 312]: 'For Descartes the laws of motion were deducible from the perfection of God, whilst for Eddington they are deducible from the peculiarities of the human mind.' Moreover, 'For both philosophers the experiments are rather a concession to our muddle-headedness and lack of insight.'

Eddington's ambitious project of reconstructing fundamental physics amounted to a theory of everything. The lofty goal was to deduce all laws and, ultimately, all phenomena of nature from epistemological considerations alone, thereby establishing physics on an a priori basis where empirical facts were in principle irrelevant. In RTPE [19, pp. 3–5] he expressed his ambition as follows:

> It should be possible to judge whether the mathematical treatment and solutions are correct, without turning up the answer in the book of nature. My task is to show that our theoretical resources are sufficient and our methods powerful enough to calculate the constants exactly— so that the observational test will be the same kind of perfunctory verification that we apply sometimes to theorems in geometry. I think it will be found that the theory is purely deductive, being based on epistemological principles and not on physical hypotheses.

At the end of the book [19, p. 327] he returned to the theme, now describing his aim in analogy with Laplace's omniscient intelligence or demon appearing in the *Exposition du Système du Monde* from 1796. However, there was the difference that Eddington's demon was essentially human in so far that it had a complete knowledge of our mental faculties. He wrote:

> An intelligence, unacquainted with our universe, but acquainted with the system of thought by which the human mind interprets to itself the content of its sensory experience, should be able to attain all the knowledge of physics that we have attained by experiment. He would not deduce the particular events or objects of our experience, but he would deduce the generalisations we have based on them. For example, he would infer the existence and properties of radium, but not the dimensions of the earth.

Likewise, the intelligence would deduce the exact value of the cosmical number (as Eddington had done) but not, presumably, the value of Avogadro's number.

Eddington's proud declaration of an aprioristic, non-empirical physics was a double-edged sword. On the one hand, it promised a final theory of fundamental physics in which the laws and constants could not conceivably be violated by experiment. On the other hand, the lack of ordinary empirical testability was also the Achilles-heel of the theory and a main reason why most physicists refused taking it seriously. Eddington was himself somewhat ambivalent with regard to testable predictions and did not always follow his rationalist rhetoric. He could not and did not afford the luxury of ignoring experiments altogether, but tended to accept them only when they agreed with his calculations. If this was not the case he consistently and often arrogantly explained away the disagreement by putting the blame on the measurements rather than the theory. Generally he was unwilling to let a conflict between a beautiful theory and empirical data ruin the theory. 'We should not,' Eddington [18, p. 211] wrote, 'put overmuch confidence in the observational results that are put forward until they have been confirmed by theory.'

## 3.6 Cosmo-Physics

In order to understand Eddington's 'flight of rationalist fancy' [35, pp. 168–191] it is important to consider it in its proper historical context. If seen within the British tradition of so-called cosmo-physics his ambitious research project was not quite as extreme as one would otherwise judge it. A flight of rationalist fancy it was, but in the 1930s there were other fancies of the same or nearly the same scale. To put it briefly, there existed in Britain in the 1930s a fairly strong intellectual and scientific tradition that in general can be characterised as anti-empirical and pro-rationalist, although in some cases the rationalism was blended with heavy doses of idealism. According to scientists associated with this attempt to rethink the foundation of physical science, physics was inextricably linked to cosmology. In their vision of a future fundamental physics, pure thought counted more heavily than experiment and observation. The leading cosmo-physicists of the interwar period, or as Dingle [4] misleadingly called them, the 'new Aristotelians,' were Eddington and E. A. Milne, but also Dirac, James Jeans and several other scientists held views of a roughly similar kind [27].

Although the world system of Milne, a brilliant Oxford astrophysicist and cosmologist, was quite different from the one of Eddington, on the methodological level Milne's system shared the rationalism and deductivism that characterized Eddington's system. Among other things, the two natural philosophers had in common that their ideas about the universe—or about fundamental physics—gave high priority to mathematical reasoning and correspondingly low priority to empirical facts. Milne, much like Eddington, claimed that the laws of physics could ultimately be obtained from pure reasoning and processes of inference. His aim was to get rid of all contingencies by turning the laws of nature into statements no more arbitrary than mathematical theorems. As Milne [32, pp. 10–12] put it: 'Just as the mathematician never needs to ask whether a constructed geometry is true, so there is no need to ask whether our kinematical and dynamical theorems are true. It is sufficient that they are free from contradictions.'

Despite the undeniable methodological affinity between the views of Milne and Eddington, the Cambridge professor insisted that his ideas were wholly different from those of his colleague in Oxford. Eddington [20] either ignored Milne's theory or he criticized it as contrived and even 'perverted from the start.'

Dirac's cosmological theory based on the $G(t)$ assumption was directly inspired by the ideas of Milne and Eddington. His more general view about fundamental physics included the claim of a pre-established harmony between mathematics and physics, or what he saw as an inherent mathematical quality in nature. By the late 1930s Dirac reached the conclusion that ultimately physics and pure mathematics would merge into one single branch of sublime knowledge. In his James Scott Lecture delivered in Edinburgh in early 1939, he suggested that in the physics of the future there would be no contingent quantities at all. Even the number and initial conditions of elementary particles, and also the fundamental constants of nature, must be subjects to calculation. Dirac [6, p. 129] proposed yet another version of Laplace's intelligence:

> It would mean the existence of a scheme in which the whole of the description of the universe has its mathematical counterpart, and we must assume that a person with a complete knowledge of mathematics could deduce, not only astronomical data, but also all the historical events that take place in the world, even the most trivial ones. The scheme could not be subject to the principle of simplicity since it would have to be extremely complicated, but it may well be subject to the principle of mathematical beauty.

Note that Dirac's version included even 'the most trivial' events in the world. This was not Eddington's view, for he believed that contingent facts—those 'which distinguish the actual universe from all other possible universes obeying the same laws'—were 'born continually as the universe follows its unpredictable course' [21, p. 64]. Another major difference between the two natural philosophers was Dirac's belief that the laws of physics, contrary to the rules of mathematics, are chosen by nature herself. This evidently contradicted Eddington's basic claim that physical knowledge is wholly founded on epistemological considerations.

## 3.7   Nature as a Product of the Mind

Although Eddington's project had elements in common with the ideas of Milne and other cosmo-physicists of the period, it was unique in the way he interpreted it philosophically. As mentioned in Sect. 3.2, Eddington was convinced that the laws of nature were subjective rather than objective. The laws, he maintained, were not summary expressions of regularities in an external world, but essentially the constructions of the physicists. This also applied to the fundamental constants of nature. Eddington [21, p. 57] characterized his main exposition of philosophy of physics, *The Philosophy of Physical Science*, as 'a philosophy of subjective natural law.' Referring to the cosmical number $N^*$, elsewhere in the book [p. 60] he explained that 'the influence of the sensory equipment with which we observe, and the intellectual equipment with which we formulate the results of observation as knowledge, is so far-reaching that by itself it decides the number of particles into which the matter of the universe appears to be divided.'

In agreement with his religious belief as a Quaker, Eddington deeply believed in an open or spiritual world separate from the one we have empirical access to. He often pointed out that physics is restricted to a small part of what we experience in a wider sense, namely what can be expressed quantitatively or metrically. 'Within the whole domain of experience [only] a selected portion is capable of that exact representation which is requisite for development by the scientific method,' he wrote in *The Nature of the Physical World* [13, p. 275]. Far from wanting physics to expand its power to the spiritual or non-metrical world, Eddington found it preposterous to believe that this world could be ruled by laws like those known from physics or astronomy [9, p. 131]. Given that his theory was limited to the metrical world, it was not really a theory of everything.

A key element in Eddington's epistemology was what he referred to as 'selective subjectivism.' With this term he meant that it is the mind which determines the nature and extent of what we think of as the external world. We force the phenomena into forms that reflect the observer's intellectual equipment and the instrument he uses, much like the bandit Procrustes from Greek mythology. The physicist, Eddington [19, p. 328] wrote, 'might be likened to a scientific Procrustes, whose anthropological studies of the stature of travellers reveal the dimensions of the bed in which he has compelled them to sleep.' As a result of the selective subjectivism, 'what we comprehend about the universe is precisely that which we put into the universe to make it comprehensible.'

Eddington's anthropomorphic and constructivist view of laws of nature was related to the conventionalist view of scientists such as Karl Pearson and Henri Poincaré, only did it go much farther. Eddington [18, p. 1] recognized the similarity to the view of the great French mathematician, from whose book *The Value of Science* he approvingly quoted: 'Does the harmony which human intelligence thinks it discovers in Nature exist apart from such intelligence? Assuredly no. A reality completely independent of the spirit that conceives it, sees or feels it, is an impossibility.'

The basic idea of the human mind as an active part in the acquisition of knowledge in the physical sciences, or even as the generator of the fabric of the cosmos, was not a result of Eddington's fundamental theory developed in the 1930s. More than a decade earlier, in the semi-popular *Space, Time and Gravitation* [10, p. 200], he wrote: 'We have found that where science has progressed the farthest, the mind has bur regained from nature that which the mind has put into nature. We have found a strange foot-print on the shores of the unknown. We have devised profound theories, one after another, to account for its origin. At last, we have succeeded reconstructing the creature that made the foot-print. And Lo! It is our own.'

## 3.8   Quantum Objections

Eddington's numerological and philosophical approach to fundamental physics attracted much attention among British scientists, philosophers and social critics in particular. The general attitude was critical and sometimes dismissive, as illustrated by the philosopher Susan Stebbing [36], who in a detailed review took Eddington to task for what she considered his naïve philosophical views. He was, she said, a great scientist but an incompetent philosopher. Leading theoretical physicists preferred to ignore the British astronomer-philosopher's excursion into unified physics. Many may have shared the view of Wolfgang Pauli, who in a letter of 1929 described Eddington's ideas as 'complete nonsense' and 'romantic poetry, not physics' [30, p. 109]. Pauli referred specifically to Eddington's identification of the fine-structure constant $\alpha$ with the number $1/136$. A main reason for the generally unsympathetic response to Eddington's theory in the physics community was his unorthodox use and understanding of quantum mechanics. I shall limit myself to some facets of this issue.

Eddington's critique of the standards employed in quantum mechanics generally fell on deaf ears among experts in the field. One of the few exceptions was a paper of 1942 in which Dirac, together with Rudolf Peierls and Maurice Pryce, politely but seriously criticized Eddington's 'confused' use of relativistic quantum mechanics. As the three physicists pointed out: 'Eddington's system of mechanics is in many important respects completely different from quantum mechanics [and] he occasionally makes use of concepts which have no place there' [7, p. 193]. The sharp difference between Eddington's quantum-cosmological theory and established quantum mechanics had earlier been highlighted at a conference on 'New Theories in Physics' held in Warsaw and Cracow in June 1938. On this occasion Eddington [22] gave a lecture in front of some of the peers of orthodox quantum mechanics, including Niels Bohr, Léon Rosenfeld, Louis de Broglie, Oskar Klein, Hendrik Kramers, John von Neuman, George Gamow and Eugene Wigner (Fig. 3.2).

None of the distinguished quantum physicists could recognize in Eddington's presentation what they knew as quantum mechanics. Kramers commented: 'When listening to Prof. Eddington's interesting paper, I had the impression that it concerned another quantum theory, in which we do not find the formulae ordinarily used, but where we find many things in contradiction with the ordinary theory.' In the proceedings of the Polish conference one gets a clear impression of how Eddington on the one hand, and Bohr and his allies on the other, failed to communicate. It was one

**Fig. 3.2** The Warsaw-Cracow meeting of 1938. Eddington sits alone on the *first row. Second row, second* and *fourth from left*, Gamow and Rosenfeld; *third row from right*, Klein, Wigner, Margrethe Bohr and Niels Bohr; *fourth row*, *second from left*, Charles Darwin. *Source* BCW

paradigm challenging another, apparently incommensurable paradigm. The attempt to create a dialogue between Bohr and Eddington led to nothing. According to the proceedings, Bohr 'thought that the whole manner of approaching the problem which Professor Eddington had taken was very different from the quantum point of view.' And Eddington, on his side, stated that 'he could not understand the attitude of Prof. Bohr.' He somewhat lamely responded that he just tried to do for quantum mechanics what Einstein had done for classical, non-quantum mechanics.

Eddington realized that he was scientifically isolated, yet he felt that the lack of appreciation of his ideas was undeserved and would change in the future. Near the end of his life he confided in a letter to Dingle [5, p. 247] that he was perplexed that physicists almost universally found his theory to be obscure. He defended himself: 'I cannot seriously believe that I ever attain the obscurity that Dirac does. But in the case of Einstein and Dirac people have thought it worth while to penetrate the obscurity. I believe they will understand me all right when they realize that they have got to do so.' Although the large majority of physicists dismissed Eddington's theory there was one notable exception, namely Erwin Schrödinger. In papers from the late 1930s the father of wave mechanics enthusiastically supported Eddington's quantum-cosmological theory [27, 33]. Yet, his enthusiasm cooled as it dawned upon him that the theory could not be expressed in a language accessible to the physicists. In an essay originally written in 1940 but only published much later, Schrödinger [34, p. 73] admitted that an important part of Eddington's theory 'is beyond my understanding.' Still today this is the general verdict of Eddington's grand attempt to establish fundamental physics on an entirely new basis.

# References

1. Barrow, J.D.: The Constants of Nature: From Alpha to Omega. Jonathan Cape, London (2004)
2. Broad, C.: Discussion of Sir Arthur Edington's 'the philosophy of physical science'. Philosophy **15**, 301–312 (1940)
3. Davis, B.: A suggestive relation between the gravitational constant and the constants of the ether. Science **19**, 928–929 (1904)
4. Dingle, H.: Modern Aristotelianism. Nature **139**, 784–786 (1937)
5. Dingle, H.: Sir Arthur Eddington, O.M., F.R.S. Proc. Phys. Soc. **57**, 244–249 (1945)
6. Dirac, P.A.M.: The relation between mathematics and physics. Proc. Phys. Soc. (Edinburgh) **59**, 122–129 (1939)
7. Dirac, P.A.M., Peierls, R., Pryce, M.: On Lorentz invariance in the quantum theory. Proc. Camb. Philos. Soc. **38**, 193–200 (1942)
8. Durham, I.: Sir Arthur Eddington and the foundation of modern physics. arXiv:quant-ph/0603146 (2006)
9. Douglas, A.V.: The Life of Arthur Stanley Eddington. Thomas Nelson & Sons, London (1956)
10. Eddington, A.S.: Space, Time and Gravitation: An Outline of the General Relativity Theory. Cambridge University Press, Cambridge (1920a)
11. Eddington, A.S.: The meaning of matter and the laws of nature according to the theory of relativity. Mind **120**, 145–158 (1920b)
12. Eddington, A.S.: The Mathematical Theory of Relativity. Cambridge University Press, Cambridge (1923)

13. Eddington, A.S.: The Nature of the Physical World. Cambridge University Press, Cambridge (1928)
14. Eddington, A.S.: The charge of an electron. Proc. R. Soc. A **122**, 358–369 (1929)
15. Eddington, A.S.: On the value of the cosmical constant. Proc. R. Soc. A **133**, 605–615 (1931)
16. Eddington, A.S.: The theory of electric charge. Proc. R. Soc. A **138**, 17–41 (1932)
17. Eddington, A.S.: The Expanding Universe. Cambridge University Press, Cambridge (1933)
18. Eddington, A.S.: New Pathways in Science. Cambridge University Press, Cambridge (1935)
19. Eddington, A.S.: Relativity Theory of Protons and Electrons. Cambridge University Press, Cambridge (1936)
20. Eddington, A.S.: The cosmological controversy. Sci. Prog. **34**, 225–236 (1939a)
21. Eddington, A.S.: The Philosophy of Physical Science. Cambridge University Press, Cambridge (1939b)
22. Edington, A.S.: Cosmological applications of the theory of quanta. In: New Theories in Physics, pp. 173–205. Scientific Collection, Warsaw (1939c)
23. Eddington, A.S.: Fundamental Theory. Cambridge University Press, Cambridge (1946)
24. Ellis, G.F.R., Uzan, J.-P.: 'c' is the speed of light, isn't it? Am. J. Phys. **73**, 240–247 (2006)
25. French, S.R.: Scribbling on the blank sheet: Eddington's structuralist conception of objects. Stud. Hist. Philos. Mod. Phys. **34**, 227–259 (2003)
26. Kilmister, C.W.: Eddington's Search for a Fundamental Theory: A Key to the Universe. Cambridge University Press, Cambridge (1994)
27. Kragh, H.: Cosmo-physics in the thirties: towards a history of Dirac cosmology. Hist. Stud. Phys. Sci. **13**, 69–108 (1982)
28. Kragh, H.: Magic number: a partial history of the fine-structure constant. Arch. Hist. Exact Sci. **57**, 395–431 (2003)
29. Kragh, H.: Cosmologies with varying speed of light: a historical perspective. Stud. Hist. Philos. Mod. Phys. **37**, 726–737 (2006)
30. Kragh, H.: Higher Speculations: Grand Theories and Failed Revolutions in Physics and Cosmology. Oxford University, Oxford (2011)
31. Kragh, H., Reeves, S.: The quantum pioneers: a bibliometric study. Physis **28**, 905–921 (1991)
32. Milne, E.A.: Kinematic Relativity. Clarendon Press, Oxford (1948)
33. Rüger, A.: Atomism from cosmology: Erwin Schrödinger's work on wave mechanics and space-time structure. Hist. Stud. Phys. Sci. **18**, 377–401 (1988)
34. Schrödinger, E.: The general theory of relativity and wave mechanics. In: Scientific Papers Presented to Max Born, pp. 65–80. Oliver and Boyd, Edinburgh (1953)
35. Singh, J.: Great Ideas and Theories of Modern Cosmology. Dover Publications, New York (1970)
36. Stebbing, L.S.: Philosophy and the Physicists. Methuen, London (1937)
37. Wesson, P.: On the re-emergence of Eddington's philosophy of science. Observatory **120**, 59–62 (2000)
38. Weyl, H.: Eine neue Erweiterung der Relativitätstheorie. Ann. Phys. **59**, 101–133 (1919)
39. Whittaker, E.T.: Review of RTPE. Observatory **60**, 14–23 (1937)

# Chapter 4
# All Possible Perspectives: A (Partial) Defence of Eddington's Physics

**Dean Rickles**

## 4.1  Eddington's Tarnished Reputation

> We may thus look on the universe as a symphony played on seven primitive constants as music is played on the seven notes of a scale.
> Eddington, *New Pathways in Science*, [5], p. 231.

*Time Magazine* referred to Eddington as "[o]ne of mankind's most reassuring cosmic thinkers" (Monday, Dec. 4th, 1944). Yet by the time of his death Eddington had lost the respect of a large sector of those working within the physical sciences. Indeed, this very 'cosmicity,' reaching its apogee in his unfinished book *Fundamental Theory*, is precisely what triggered Eddington's downfall in the minds of many physicists and mathematicians, viewed by them as a journey into numerology and mysticism. His approach to science became the butt of many jokes. Richard Tolman spoke of Eddington's method as pulling "rabbits of physical principle out of the hat of epistemology". Harold Jeffreys wrote that "There is probably a great deal of good physics in Eddington's *Fundamental Theory*, but at present this is so overlain by bad epistemology that it is very hard to see what it is" ([10], p. 175).

At the root of the problems was Eddington's claim to be able to deduce pure numbers of physical significance from aspects of how observers are constrained to gain knowledge about and interact with the world—an aspect curiously close to Wheeler's participatory approach to the meaning of quantum theory. Hermann Bondi wrote that "[t]he wealth of startling numerical results of his theory which offered explanations for most of the pure numbers revealed by observation seemed to act mainly as a deterrent to the study of his work" ([3], p. 158). In this article I will follow Edmund Whittaker's view that "too much notice has been taken of the numbers, and too little of the general principles" ([16], p. 24).

D. Rickles (✉)
University of Sydney, Sydney, NSW, Australia
e-mail: dean.rickles@sydney.edu.au

© Springer International Publishing Switzerland 2017
I.T. Durham and D. Rickles (eds.), *Information and Interaction*,
The Frontiers Collection, DOI 10.1007/978-3-319-43760-6_4

Though there are indeed very many things Eddington was wrong about, I hope to show that some aspects of his general framework for thinking about physical theories are defensible and relevant today. In particular, his analysis of measurement and observables in generally relativistic theories foreshadowed, by more than half a century, the analyses given by the likes of Carlo Rovelli (in his framework of partial and complete observables)—it is also highly likely that Eddington's analyses influenced those of Bergmann that were direct precursors to the treatment of Rovelli et al. Moreover, Eddington provides a more cogent physical interpretation, grounded in an irreducibly structural understanding of the physical observables.

## 4.2    Eddingtonian *A Priori* and *A Posteriori*: *Quis Custodiet Ipsos Custodes*?

In his book *The View from Nowhere*, Thomas Nagel argued that though "[a]ll of our thoughts must have a form which makes them accessible from a human perspective ... that doesn't mean they are all about our point of view or the world's relation to it" ([12], p. 102). For Eddington, in the case of our scientific knowledge about physics, our point of view is central. When we give a treatment of quantities in physical theories, we must incorporate two basic elements: on the one hand, of course, there are the physical objects whose properties we are attempting to model, and on the other there is a mathematical framework used to represent the quantities. In order to achieve an 'objective' account, we need to somehow extract the human contribution that is employed in the mathematical framework (which function as a kind of generalised set of co-ordinates). Eddington believed that the mathematical representation of physical systems (the *physical universe*, as he called it) was deeply entangled with our perspective as observers in a way that was ineliminable for all practical purposes. The way we are constrained to perform measurements and make observations to determine features of the world, largely infects the nature of the mathematical framework used.

At the root of the general distaste for Eddington's later proposal was what he himself referred to as "the family skeleton": the possibility of *a priori* knowledge about the physical world (see [7, p. 24]).[1] He defines this as "knowledge which we have of the physical universe prior to actual observation of it". This is already off more orthodox definitions of *a priori* knowledge, since it points to some future or possible *observation*. That this is Eddington's intended meaning is borne out in the next paragraph, in which he remarks that "*a priori* knowledge is prior to the carrying out of observations, but not prior to the development of a plan of observation" (ibid.). Thus, in Eddington's notion of *a priori* knowledge, observation and observability remain central components. He goes on to explicitly deny that such knowledge is "independent of observational experience": Eddingtonian *a priori* refers to the process

---

[1] Edmund Whittaker refers to the problem of "the respective shares of reason and sense-perception in the discovery of the laws of nature" ([15], p. 185).

of observation and measurement. He has a particularly nice example to make his idiosyncratic[2] notion clear:

> A valuer may arrive at the generalisation *a posteriori* that no article in a certain house is worth more than sixpence; the same generalisation might also have been reached *a priori* by noticing that the owner furnished it from Woolworth's.[3] The observer is called upon to supply the furniture of the mansion of science. The priorist by watching his method of obtaining the furniture may anticipate some of the conclusions which the posteriorist will reach by inspecting the furniture.

Let us lay out Eddington's understandings as starkly as possible, to draw out the fact that they are non-standard:

A Priori: epistemological or *a priori* knowledge is prior to the carrying out of observations, but not prior to the development of a plan of observation ([7], p. 24).

A Posteriori: Result of observation or measurement.

Clearly, and unusually, both invoke observation and measurement. To have *a priori* knowledge is to have in mind the potential performance of some physical observation. This exposes the themes I'd like to pursue in the remainder of this paper. Firstly, it makes it very clear that this is not *a priori* knowledge in the usual sense. Secondly, the sense in which it differs highlights what I think is absolutely crucial in making sense of (and providing an adequate defence of) Eddington's entire scheme. This difference centres on the concept of 'observability': Eddington's *a priori* scientist does not introspectively analyse the contents of his head, but instead himself observes *observers* and the necessary conditions demanded by the construction of observables. In other words, the subject of Eddington's project is (scientific) observation itself. Given his way of understanding fundamental physics, the discoveries encoded in the most general laws are really discoveries about the processes of measurement and observation. He has several colourful passages that make this point:

> We have found a strange footprint on the shores of the unknown. We have devised profound theories, one after another, to account for its origins. At last, we have succeeded in reconstructing the creature that made the footprint. And lo! It is our own.

> The physicist might be likened to a scientific Procrustes, whose anthropological studies of the stature of travellers reveal the dimensions of the bed in which he has compelled them to sleep ([6], pp. 328–329).

That is, what were thought to be claims about *objective* facts, independent of observers, really arise from features of those very observers. As he points out, "whatever is accounted for epistemologically [by observing observers–DR] is *ipso facto* subjective; it is demolished as part of the objective world" ([7], p. 59).[4]

---

[2]In truth, it is simply a bad use of terminology, bound to confuse readers acquainted with standard discussions of the *a priori*.

[3]Note that this was written in the days when Woolworth's was solely a threepence and sixpence arcade.

[4]We might note, that Wheeler's 'it from bit' notion, though very similar, denies the lack of objectivity of such knowledge, promoting the information (as encoded in apparatus-elicited answers to 'yes-no'

## 4.3  Going Soft on Truth?

The phrase "going soft on truth" comes from Michael Redhead's Tarner Lectures ([14], p. 15), where he uses it in the context of a dismissal of subjectivism. Eddington's own Tarner Lectures (published as *The Philosophy of Physical Science*) describes a position that he labels "selective subjectivism". In tandem with each of his technical books, *Mathematical Theory of Relativity*, *Relativity Theory of Protons and Electrons*, and *Fundamental Theory*, Eddington wrote a 'popular' account for a general audience: *The Nature of the Physical World* [NPW], *New Pathways in Science* [NPS], and *The Philosophy of Physical Science* [PPS]. These books chart a move towards greater subjectivism with each 'advance' based, he argued, on a switch "in science itself"[5]:

NPW: "the nature of the physical universe with applications to knowledge"

NPS: "the structure is the object of our search" = "a physical world which will give a shadow performance of the drama enacted in the world of experience"

PPS: "the nature of knowledge with applications to the physical universe" and "the universe which physical science describes is partially subjective" (PPS, p. 26)

Does the final point, in PPS, amount to going soft on truth? Yes, in a sense: it's truth within a system of physical knowledge. Eddington compares physical knowledge to the kinds of true statements one can make about Pickwick, within the context of that world. The methods of physical science yield knowledge of "a detailed description of a world" which Eddington calls "the physical universe". Then "scientific epistemology" is a sub-branch of epistemology applied to this description (and the nature and status of the physical universe). The physical universe, like our knowledge of it, is 'non-stationary'.

The fundamental laws are (partially) subjective, while particular events are objective—Eddington claimed to be able to deduce generalities [invariants], not particular events:

> We cannot...predict the result of the measurement from our *a priori* knowledge of how the measurement is going to be made; but it does not seem unlikely that we should be able to predict certain general properties that the resulting measurements will have.

He writes that "[p]hysical science may be defined as 'the systematisation of knowledge obtained by measurement'. It is a convention that this knowledge shall be

---

(Footnote 4 continued)

questions) to the most objective facts of the world. I have a feeling that Eddington would not have disagreed, however, and was using 'subjective' and 'objective' in an idiosyncratic manner: if subjective for Eddington means 'relative to observers,' then I think Wheeler and Eddington are on the same page. The real difference in their positions, however, stems from the fact that quantum mechanics plays the central (essential) role in Wheeler's approach, whereas in Eddington's it is considerations of a more general sort, having to do with measurables and observables that are crucial.

[5]He describes PPS as a study of "principles of philosophic thought associated with the modern advances of physical science." He argued that "philosophic truth should be reached by the same method of progressive advance" as in the sciences.

formulated as a description of a world—called 'the physical universe'" ([9], p. 268). This picture, grounded in measurement, is deeply entangled with the process of observation. Eddington has a famous analogy between nets and fishing on the one hand and our sensory/cognitive faculties and data gathering methods on the other. The idea is that the throwing of a net is analogous to our performing of experiments. The net itself is analogous to the sense organs and cognitive equipment of observers. The net's catch is analogous to physical knowledge gained via our senses and processed via our cognitive equipment. In the case of the net, one can readily imagine a situation in which a fisherman throws a net with a mesh size of, say, 2 cm$^2$. Suppose the fisherman brings in a large catch. It is clear that every specimen in the catch will be greater than 2 cm in size. We would not, however, allow the validity generalisation that all fish are greater than 2 cm$^2$ since the net constrains any fish caught to be greater than this. There might well be smaller fish that the net is unable to hold. Moreover, we could have concluded that any fish caught will be greater than 2 cm$^2$ without throwing the net in the first place: features of the catch were predetermined by features of the net.

As with nets, so with our access to the world argues Eddington: features of physical knowledge will be predetermined by features of our makeup. The fundamental laws tell us about our method of observation. They are "consequence[s] sof the conceptual frame of thought into which our observational knowledge is forced by our method of formulating it". He labels this position "selective subjectivism". This opens up the possibility of a *testable* hypothesis: we should be able to deduce laws from a close analysis of our methods for making measurements and observations, without engaging with the physical systems themselves, whose details will be unimportant given the general nature of the laws under consideration. This is where what is considered 'the crazy stuff'[6] originates: if laws are epistemological (that is, Eddingtonian *a priori*) in origin, then we should be able to deduce them from features of observers and observation. The crazy stuff, then, was Eddington's way of justifying his general 'selective subjectivist' position—a kind of empirical test of Kantianism.

## 4.4  Eliminating All Possible Perspectives: The Epistemological Purge?

In his *View from Nowhere*, Thomas Nagel presents the following 'Perspectival Problem': "how to combine the perspective of a particular person inside the world with an objective view of that same world, the person and his viewpoint included?" ([12], p. 3). Redhead, in his Tarner Lectures again, seemingly in answer to Nagel, writes: "We, as observers, each have a subjective perspective in the world. The aim of

---

[6]Freeman Dyson once wrote: "The crazy ideas about calculating the constants of nature from first principles had roughly the same place in Eddington's life that alchemy had in Newton's. ... Unfortunately Eddington spoiled his reputation by publishing his crazy ideas; in this respect Newton was wiser." [Applications of Group Theory in Particle Physics, SIAM Review, 8(1): 1].

science involves patching these perspectives together, and abstracting an 'objective' viewpoint". Now, Eddington agrees that, 'what he calls 'personal' subjectivity' can be removed:

> Since physical knowledge must in all cases be an assertion of the results of observation (actual or hypothetical), we cannot avoid setting up a dummy observer; and the observations which he is supposed to make are subjectively affected by his position, velocity and acceleration. The nearest we can get to a non-subjective, but nevertheless observational, view is to have before us the reports of all possible dummy observers, and pass in our minds so rapidly from one to another that we identify ourselves, as it were, with all the dummy observers at once. To achieve this we need a revolving brain ([7], pp. 86–87).

Of course, this will be achieved by adopting a tensorial formalism, encoding all possible (personal subjective) perspectives. Picking a gauge (coordinate representation) is then tantamount to picking a specific dummy observer. However, Eddington argues that this leaves a residual 'generic' subjectivity, harking from deeper aspects of our sensory and cognitive equipment, and the way observers such as ourselves are constrained to interact with the world. To eliminate this, he tells us, would require symbolisation of "knowledge as it would be apprehended by all possible types of intellect at once" ([7], p. 87). Hence, there is an ineliminable core of subjectivity in all of our theorising about the world.[7]

## 4.5  Postulates of Impotence

Going back to the example of the net, we saw that net-analysis could tell you something about what you *couldn't* catch. By analogy, epistemological analysis could tell you something about what you couldn't *observe*. These take the form of laws of a qualitative, negative character; they are what Whittaker calls "Postulates of

---

[7]I note that Poincaré presented a very similar view: "Since the enunciation of our laws may vary with the conventions that we adopt, since these conventions may modify even the natural relations of these laws, is there in the manifold of these laws something independent of these conventions and which may, so to speak, play the role of universal invariant? For instance, the fiction has been introduced of beings who, having been educated in a world different from ours, would have been led to create a non-Euclidean geometry. If these beings were afterwards suddenly transported into our world, they would observe the same laws as we, but they would enunciate them in an entirely different way. In truth there would still be something in common between the two enunciations, but this is because the beings do not yet differ enough from us. Beings still more strange may be imagined, and the part common to the two systems of enunciations will shrink more and more. Will it thus shrink in convergence to zero, or will there remain an irreducible residue which will then be the universal invariant sought?" ([13], p. 334). He seemed to provide a broadly similar conclusion to Eddington, based on his notion of 'convention': "What now is the nature of this invariant it is easy to understand, and a word will suffice us. The invariant laws are the relations between the crude facts, while the relations between the 'scientific facts' remain always dependent on certain conventions" (ibid, p. 336).

Impotence".[8] Examples are: the impossibility of perpetual motion machines; the equivalence principle; the impossibility of detecting (global) uniform translatory motion from within the system; and the impossibility of measuring precise simultaneous values of position and momentum. This fits Eddington's characterisation:

> One consequence of the new outlook . . . is that a law of Nature has a compulsory character, and is not merely an empirical regularity which we have hitherto found to be fulfilled in our limited experience. Conflict with the laws of Nature is impossible. "Impossible" you may say "is a word which no genuine scientist should use". Very well; let us substitute another—which makes the laws no less compulsory. Conflict with the laws of Nature is unobservable.

Postulates of Impotence are epistemologically curious, being neither direct inferences from experiments, nor theorems of pure mathematics, nor creations of the intellect. Instead, they are convictions about prohibitions on certain kinds of observation. These are precisely the kinds of qualitative statement that Eddington considered *a priori*—this has often been ignored since it seems to stretch the notion of *a prior* too far. He makes syllogistic derivations of scientific claims from such Postulates of Impotence "without turning up the answer in the book of nature".[9] Eddington restricted his attention to the dimensionless constants alone, of which he claimed to be able to deduce their exact values, with "the observational test [being] the same kind of perfunctory verification that we apply sometimes to theorems in geometry." ([16], pp. 3–4).

## 4.6   From Pure Numbers to Eddington's Principle

We have to be very clear on what Eddington thought he could derive *a priori*. A criticism frequently levelled at him was that he assumes all sorts of physical facts that themselves have their origin in experience. This is perfectly true, and he nowhere denies this: he is quite prepared to take on, as building materials for his project, any qualitative facts about the physical universe gained through observation. The equivalence principle and Pauli's exclusion principle are both examples of data used by Eddington but not derived. His deductions were instead restricted to certain pure numbers: he believed he could compute the exact values of those numbers that are constants of science. Hence though he assumed a great deal from physics, he

---

[8]Whittaker points out that many qualitative features can shift during theory change, so since the quantitative assertions are based on qualitative features in Eddington's scheme, they too must be incomplete. Postulates of Impotence are supposed to be more stable.

[9]Of course, the complaint would be that he sneaked a peek at the various pages of the book in the postulates of impotence he employs. However, this is by the by. Eddington admits to using such qualitative assertions. What he doesn't do is find *quantitative* values of constants in this way: these he formally deduces from his qualitative assertions.

does not assume such constants. Thus, Eddington's project to derive the fundamental constants and laws of physics is rooted in *pure numbers*.[10]

One can construct pure numbers from 'impure' numbers of physics (thus eliminating arbitrary units of MLT), and such numbers play a central role in physics and are crucial in Eddington's scheme, since he has these in his sights as computable from epistemological principles.[11] The fine-structure constant (representing the coupling strength for electromagnetic interactions) is just such a construct. From $e$ (electric charge), $c$ (speed of light *in vacuo*), and $h$ (unreduced Planck constant) one can form: $hc/2\pi e^2$. Eddington derives a fixed value of 137 for this constant from theory.[12] Eddington noticed (as did others) that in fundamental physics and cosmology the pure numbers are not randomly distributed about the real axis, but instead cluster around three points:

1. Unity: the fine-structure constant; proton-electron mass ratio
2. $10^{39}$: the baryon-universe radius ratio; ratio of gravitation force to electric force [Force constant]
3. $10^{78}$: the "cosmic number" enumerating the number of protons in the universe.

Eddington claimed that these can be seen as advancing in order of fundamentality: the square root relation relates $10^{39}$–$10^{78}$.

Very early on, for example in his book *Space, Time, and Gravitation* (1920), Eddington focused on the importance of pure and impure numbers and their relation to experiment, observation, and reason. For example, he writes:

> [A]ny physical quantity, such as length, mass, force, etc., which is not a pure number, can only be defined as the result arrived at by conducting a physical experiment according to specified rules.

In other words, dimensionful quantities cannot be derived purely theoretically. His later work would attack the converse: dimensionless quantities *can* be arrived at both by experiment *and* theory. Dimensionful quantities can only be made sense of when coupled to some definition specifying units and so on. To speak of a mass, for example, demands that we have in mind some way of measuring mass, involving a standard.

Edmund Whittaker claimed that a new principle of philosophy of science is the central theme of *Fundamental Theory*. It was based on the distinction between quantitative [e.g. fine structure constant] and qualitative aspects of physics [e.g. observable symmetries]:

---

[10]Numbers that's don't depend on units of Mass, Length, and Time are said to be pure (of dimension 1 in dimensional analysis). So, for example, the ratio of proton mass and electron mass is pure since it is invariant under alterations in the system of units.

[11]This is very similar to the way that one forms a gauge-invariant physical quantity from a pair of gauge-variant quantities: it is a form of gauge-invariance.

[12]He famously initially derived 136, but later modified his reasoning as a result of experiment. He claimed a permutation invariance had been neglected—see below.

> All the quantitative propositions of physics, that is the exact values of the pure numbers that are constants of science, may be deduced by logical reasoning from qualitative assertions, without making any use of quantitative data derived from observation ([16], p. 3).

That is, one is not starting with zero empirical materials and building up out of pure thought elements of physical reality (a common misunderstanding). One can make use of various qualitative principles (excepting pure numbers) in the analysis and compute from them pure numbers (the quantitive features of the physical universe).

## 4.7   How Do the Deductions Work?

Let us provide an example to give some indication of how the various deductions are supposed to work. Eddington used two methods to compute the total number of particles in the universe: a posteriori and a priori (in his sense). Via the latter method he computed the number of protons in the universe to be *exactly* $N = 10^{79}$ (from $2.136 \times 2^{256}$).[13] He called this "the cosmical number". Initially, he gets the number in a fairly straightforward manner. Firstly, he is dealing with an 'Einstein world' (a finite, unbounded universe). He determines the mass and radius of this world from the recession constant $V_0$ (i.e. the increase in galactic velocities per unit distance):

$$V_0 = \frac{c}{R_0\sqrt{3}} \tag{4.1}$$

where $V_0$ is fixed by observation, which allows one to determine the universe's radius $R_0$. Given $R_0$ one then solves for the mass with:

$$\frac{\kappa M}{c^2} = \frac{1}{4}\pi R_0. \tag{4.2}$$

Next, given $M$, and the assumption that there are only protons and electrons, so all mass in the universe is constituted by protons and electrons[14] (and there is an equal number of them), and given the mass of hydrogen (made up of 1 proton and 1 electron bound by the Coulomb force), one can just divide $M$ by the hydrogen mass and multiply the result by 2 to get the total number of particles.

But Eddington thought there was a second route, starting from qualitative data (rather than quantitative data like $V_0$). The derivation involves his understanding of measurement and observables in a generally relativistic context. As will be described more fully in the next section, observables are represented by two-component entities and measurements by pairs of observables, or four-component entities (which we'll

---

[13]Expanded out the number is: 15 747 724 136 275 002 577 605 653 961 181 555 468 044 717 914 527 116 709 366 231 425 076 185 631 031 296.

[14]Of course, we now have reason to believe that baryonic matter accounts for only a very small fraction of the total mass of the universe, with the majority comprising dark matter and, more still, dark energy.

call a 'target' observable and a 'standard' observable). A measurable (given both quantum mechanics and general relativity) is represented by a quadruple wavefunction. The number of independent quadruple wavefunctions in an Einstein universe (i.e. of constant positive curvature) is identical to the number of particles. Eddington calculates $N = \frac{3}{2} \times 136 \times 2^{256}$ which, running this through *Mathematica*, yields $2.36216 \times 10^{79}$, close to the value derived from observational input. Eddington links the force constant and the cosmical number via a square root relation. $F = \frac{2}{\pi}\sqrt{N}$. Given the exactness of his computation of $N$, he can also compute $F$ exactly.

Eddington's results received very similar responses from the community to those received by string theory's claims today. Namely that they are not predictions as such, but rather accommodations at best, since the values were already known and so could function as visible targets to be hit (opening up the possibility of fitting).

## 4.8 *Real* Observability and Measurability Analysis

Eddington is perhaps the first person, after Einstein himself, to take seriously the problem of quantum gravity. His approach was firmly based in unpacking the observable content of such a theory. In their treatise on Eddington's physics, Ted Bastin and Clive Kilmister write that:

> Eddington's Fundamental Theory embodies an unorthodox attitude to the interpretation of physical measurement. This attitude regards physical theories as formulations of conditions presupposed by our experimental procedures, rather than determined, empirically, by those procedures. Eddington's method was to alter the existing structure of relativity theory in accordance with this attitude in order to make it cover quantum problems ([1], p. 59).

Eddington performed a kind of measurability analysis[15] for general relativistic theories, both classical and quantum. The conditions of observability and measurement seemingly played a central role in Eddington's 'pure deductions'. Rather than assess the specific deductions, to finish I'd like to focus on his ideas of measurement and observables. It's my contention that with respect to his analysis of observables and measurement much of value was ignored, largely because Eddington's understanding of the physical content of generally relativistic theories was too far ahead of the general understanding.

Eddington's understanding was rooted in the notion that observables in general relativity must necessarily have a relational character, linking multiple entities or quantities. We don't measure or observe things from abstract geometrical points. An observed motion is a relative motion of two physical systems. An observed location is a location of one physical system relative to another. And so on. This is fairly orthodox in current treatments of general relativity. Yet in his review of Eddington's

---

[15] As Bergmann and Smith put it, "[m]easurability analysis identifies those dynamic field variables that are susceptible to observation and measurement ("observables"), and investigates to what extent limitations inherent in their experimental determination are consistent with the uncertainties predicted by the formal theory" ([2], p. 1131).

PPS, Ernest Nagel asks: "what student of relativity will recognize the essence of that theory to consist in the fact that 'we observe only relations between physical entities'?" One wonders what Nagel would have offered in its place. Yet Eddington's analysis was more far reaching than this basic and, as I say, now standard picture. He wanted to include quantum theory too. This complicates the analysis considerably:

> [A]n observable coordinate is measured, not from an abstract geometrical origin, but from something which is involved physically in the experiment which furnishes its observed value... We must therefore distinguish between the 'physical origin' from which an observable coordinate is measured, and the 'geometrical origin' of the mathematical reference frame which is inaccessible to measurement ([8], p. 3).

As I mentioned, Eddington, like Einstein (though with more detail and rigour), was one of the first to pursue the idea that the observables of a generally relativistic theory are necessarily *relational*: a product of two entities or quantities. Eddington extended this to the quantum mechanical case, so that observables are represented as products $\phi\psi$.[16] These products are not to be viewed as ordered, in terms of ontological or formal priority, so that both are necessary for the observable to be given an ontological interpretation.[17] Such entities correspond to what Rovelli calls "complete observables" (as opposed to the individual $\psi$ and $\phi$, which are "partial observables"). Clive Kilmister puts the distinction that Rovelli would later make using his partial/complete observables in terms of 'potential' and 'actual' properties:

> [W]e may say that the so-called physical entities—the relata whose relations furnish the observables—have potential properties only, which must be expressed by a symbol $\psi$ with no observational interpretation; but two such entities in conjunction have actual properties, and $\phi\psi$ has an observational interpretation ([11], pp. 262–263).

I have argued in several places that the proper interpretation of observables in generally relativistic theories must be treated in terms of *structuralism*. The reason is clear: the relata are secondary concepts in this scenario, without any direct physical interpretation, while what Kilmister calls the 'conjunction' is primary. Note that such conjunctions are not yet *measurables*, which will require at a minimum a pair of such observables (and hence four components)—hence, a measurement is a relation that relates a pair of relations. Eddington puts the point thus: "Formally a measurement always refers to just four relata, viz. the terminals of the object relation and the comparison relation.[18] The basis of measurement is therefore a four-point element

---

[16]I save an analysis of his approach to quantum gravity for a later paper. I will note, however, that the complexities come from the fact that a physical reference frame used to localise observables, to build relative quantities such as velocities and positions, in measurements for example, will be described by a wave function.

[17]Of course, we will often adopt the practice of supposing that one or other factor is 'really' the one that *has* some property (that really moves, or is located, etc.). But this is a *façon de parler*: either could be viewed as the reference factor: "we can no more contemplate an atom without a physical universe to put it in than we can contemplate a mountain without a planet to stand it on".

[18]In *Fundamental Theory* Eddington invokes the cosmological term to function as a "comparison fluid" (an "all-pervading entity"), thus grounding a physical reference frame capable of localising observables.

of world-structure" ([6], p. 323). However, as Kilmister again makes plain, the same primacy of the conjunction[19] appears:

> When an object-body is observed in conjunction with a reference body, any measurement that we make determines characteristics which belong, neither to one body nor to the other, but to both jointly. It is, however, customary to allot these mutual characteristics to the object-body, or more defensibly to partition them between the two bodies according to some self-consistent scheme. The conceptual transfer, by which self properties are substituted for mutual properties, is a habit of thought which has been elevated into a convention, and we can scarcely do otherwise than accept in principle ([11], p. 271).

This sounds remarkably similar to Earman's slightly fumbling attempts to make ontological sense out of the gauge-invariant content of generally relativistic theories:

> Coming to grips with the ramifications of SGC indicates that we need to rethink the traditional subject-predicate ontology/ideology. I want to tentatively suggest that the gauge-invariant content of GTR is best thought of in terms of a new ontological category that I will call a coincidence occurrence. I use "occurrence" rather than "event" since the latter is traditionally conceived in subject-predicate terms, whereas coincidence occurrences lack subjects and, thus, also predicates insofar as predicates inhere in subjects; rather, a coincidence occurrence consists in the corealization of values of pairs of (non-gauge invariant) dynamical quantities. The textbook models of GTR are to be thought of as providing many-one representations of coincidence occurrences in terms of the co-occurrence of the relevant values of the pairs of quantities at a space-time point. If further pressed to give a representation-free characterization of coincidence occurrence, I have nothing to offer. . . . [M]y feeling is that spacetime theories satisfying SGC are telling us that traditional subject-predicate ontologies, whether relational or absolute, have ceased to facilitate understanding ([4], p. 16).

Here, Earman is struggling to explicate the ontological picture that results from gauge-invariance decades after Eddington's attempts. One can visualise the problem in the following diagram (Fig. 4.1). There are various levels of representational freedom in this diagram (simple passive coordinate invariance, and substantive general covariance) and the problem is to make sense of the 'thing' giving rise to the various multiple representations. As Eddington, and many others since, have argued, we will require some kind of relative quantities. However, Eddington was right in thinking that, as Earman points out, a simple decomposition into relata isn't going to work. What I call "Unruh's Decomposition Problem" sets up the problems facing such a simple decomposition rather nicely:

> [O]ne could [try to] define an instant of time by the correlation between Bryce DeWitt talking to Bill Unruh in front of a large crowd of people, and some event in the outside world one wished to measure. To do so however, one would have to express the sentence "Bryce DeWitt talking to Bill Unruh in front of a large crowd of people in terms of physical variables of the theory which is supposed to include Bryce DeWitt, Bill Unruh, and the crowd of people. However, in the type of theory we are interested in here, those physical variables are all time independent, they cannot distinguish between "Bryce DeWitt talking to Bill Unruh in front of a large crowd of people and "Bryce DeWitt and Bill Unruh and the crowd having grown old and died and rotted in their graves."

---

[19]We perhaps need a better term to describe these "conjunctions". I called them 'correlations' elsewhere, but neither adequately captures the mutual embrace and lack of ontological priority of the factors in such entities.

**Fig. 4.1** Diagram from Earman [4], with modifications by author

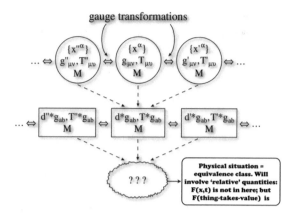

The subtle assumption is that the individual parts of the correlation, e.g. DeWitt talking, are measurable when they are not.

One cannot try to phrase the problem by saying that one measures the gauge dependent variables, and then looks for time independent correlations between them, since the gauge dependent variables are not measurable.[20]

Hence, the gauge invariant correlations (Kilmister's conjunctions and Eddington's observables) are, in a sense, irreducible. Any kind of factoring into relata reflects gauge freedom. We all too naturally fall into the trap of abstracting away reference relata from measurements and observations, inferring that whatever property of observed is possessed by an individual *object*. However, what is taken as a single object possessing some property involves a congruence of four elements.

Given that Earman is suggesting that the underlying 'thing' might force a radical revision of traditional subject-predicate ontologies, we might charitably follow Kilmister and Bastin in their assessment that "the obscurity [of Eddington's later work] is that inevitably associated with the revolutionary" ([1], p. 1).

## 4.9  Conclusion

I have tried to demonstrate in this paper that both Eddington's use of *a priori* methods, in the deduction of constants and the related analysis of measurement of observables, are the result of a kind of extreme analysis based around the question *what really happens when we perform a measurement?*[21] The apparent craziness of the deductions is significantly reduced when it is realised that the qualitative principles flowing into

---

[20]See Unruh's comments in the proceedings of the *Conference on Conceptual Issues in Quantum Gravity* (A. Ashtekar and J. Stachel, eds., Birkhäuser, 1989, p. 267).

[21]"We have to express in mathematical symbolism what we think we are doing when we measure things" ([9], p. 266).

the quantitative deductions have the status of broad physical generalities. At the root of the principles lies an analysis of measurement and observables that was decades ahead of its time, and was an interpretation of the physical content of generally relativistic theories that philosophers are only very recently rediscovering.

# References

1. Bastin, T., Kilmister, C.: Eddington's Statistical Theory. Clarendon Press, Oxford (1962)
2. Bergmann, P., Smith, G.: Measurability analysis of the linearized gravitational field. Gen. Relativ. Gravit. **14**(12), 1131–1166 (1982)
3. Bondi, H.: Cosmology. Cambridge University Press (1958)
4. Earman, J.: The implications of general covariance for the ontology and ideology of spacetime. In: Dieks, D. (ed.) The Ontology of Spacetime, pp. 3–23. Elsevier (2006)
5. Eddington, A.S.: New Pathways in Science. Cambridge University Press (1935)
6. Eddington, A.S.: Relativity Theory of Protons and Electrons. Cambridge University Press (1936)
7. Eddington, A.S.: The Philosophy of Physical Science. Cambridge University Press (1939)
8. Eddington, A.S.: The Combination of Relativity Theory and Quantum Theory. Dublin Institute for Advanced Studies (1943)
9. Eddington, A.S.: Fundamental Theory. Dublin Institute for Advanced Studies (1943)
10. Jeffreys, H.: Review of the sources of eddington's philosophy, by Herbert Dingle. Br. J. Philos. Sci. **7**(26), 174–175 (1956)
11. Kilmister, C.: Eddington's Search for a Fundamental Theory. Cambridge University Press (2005)
12. Nagel, T.: The View from Nowhere. Oxford University Press (1986)
13. Poincaré, H.: The Value of Science. The Modern Library, New York (2001)
14. Redhead, M.: From Physics to Metaphysics. Cambridge University Press (1996)
15. Whittaker, E.: From Euclid to Eddington: A Study of Conceptions of the External World. Cambridge University Press (1949)
16. Whittaker, E.: Eddington's Principle in the Philosophy of Science. Cambridge University Press (1951)

# Chapter 5
# Tracing the Arrows of Time

**Friedel Weinert**

The smart money these days is that the universe won't actually re-collapse (S. Carroll, *From Eternity to Here*, 2010, p. 349).

## 5.1 Introduction

An age-old link exists between time and cosmology. Plato, for instance, regarded the planets as the instruments of time. Even though Aristotle rejected the identification of time with celestial motion, he still viewed celestial motion as the best criterion for the measurement of the passage of time. In the Greek geocentric worldview the planets move around the 'central' Earth with perfect regularity and invariance. In fact the planets move in circular orbits around the centre. Circular orbits furnish perfect periodic regularity since the planets move at a constant speed and always stay at the same distance from the 'centre'. They therefore also provide invariance since the motion of the planet will not change through a change of perspective by transporting the observer from one circular orbit to another or from one period of time to another.

The different attitudes of Plato and Aristotle allow us to draw a distinction between the measurement of time—it requires periodic regularity and invariance—and the nature of time. Time can be measured without knowing its nature. While Plato identifies the passage of time with cosmic motion, Aristotle regards celestial motion merely as a criterion of temporal passage. He argues that we always measure motion in terms of time but never time in terms of motion. Furthermore, motion can be fast or slow but it can still be measured. But when Plato identifies the nature of time with circular motion of the planets—time is a moving image of eternity—he seems to accept that we cannot observe time and its properties directly. Rather we must use inferences from appropriate criteria. If we cannot observe time as such we must use

F. Weinert (✉)
Bradford University, Bradford, Yorkshire, UK
e-mail: F.Weinert@bradford.ac.uk

© Springer International Publishing Switzerland 2017
I.T. Durham and D. Rickles (eds.), *Information and Interaction*,
The Frontiers Collection, DOI 10.1007/978-3-319-43760-6_5

such criteria either to identify time with one of them or use these criteria as indicators, 'signposts' to infer temporal properties.

Before we turn to the modern question of the anisotropy of time, let us briefly ask whether the Greeks could have recognized a cosmological arrow of time. On the geocentric worldview there is no room for a cosmic arrow of time, since the Greek universe is closed, symmetric and eternal (at least in the superlunary sphere).[1] However, planetary motion served at least as a criterion for the passage of time in the sublunary sphere (located between the Earth and the moon), which is characterized by asymmetry and decay. Such an anisotropy could be called a 'local arrow of time'.

Arthur Eddington and John A. Wheeler both emphasised the role of inferences from 'data', 'pointer readings' or 'information' to phenomena and knowledge of physical reality. Eddington declares that 'all knowledge of physical objects is inferential' [19, p. 92]; all physical knowledge is 'hypothetico-observational knowledge' [20, Chap. I, §4]. Physical science is an inference from good observations. It arrives at structural knowledge [20, Chap. II, §3]. Referring to the Special theory of relativity, Eddington makes a distinction between 'physical time', of which we have indirect knowledge, and 'phenomenal (subjective) time' of which we possess direct knowledge through mental awareness. As we shall see, when Eddington concerns himself with the question of the arrow of time, he displays the same ambiguity as the Greeks between the 'nature' of the cosmic arrow and criteria to ground it.

J.A. Wheeler's famous catchphrase 'it from bit' is also a plea for an inference from information, gathered from various sources about the external world, to a theory of the physical world. 'Every item of the physical world has at bottom an immaterial source.' Wheeler hoped to establish the lapse of (physical) time from 3-geometry [13, 69, 70]. This chapter will however concentrate on Eddington's contribution because of the importance of the Second law of thermodynamics in his thinking.

Eddington, furthermore, distinguishes between the passage of time (clock time: how much later one event occurs after another) and the arrow of time (as the increase in randomness in accordance with the 2nd law of thermodynamics) [18, p. 99]. As Eddington sees it, at thermodynamic equilibrium $dS/dt = 0$ the arrow of time is lost—since randomness has reached its limit—but 'time is there' since 'the atoms vibrate as usual like little clocks' [18, pp. 78–79].

By contrast G. Ellis [23, §3] distinguishes between the direction of time—the evolving block universe—and the arrow of time—the local passage of time. For the purpose of this chapter I shall distinguish between the *local* arrows of time—temporal passages as indicated in biological, entropic, historical, psychological and radiative arrows—and the *cosmic* arrow of time—the global evolution of the universe [1, 11, 43]. This distinction seems justified by a number of features:

---

[1]Note that the Greeks were concerned with 'saving the appearances': they fitted their naked-eye observations of planetary motions to their preconceptions of unchanging circular orbits. Rather than allowing the observations to 'falsify' their theoretical constructions, they remained faithful to their geometric presuppositions and 'explained' away the appearance of non-constant planetary orbits. The Greeks entertained a distinction between a changing sublunary sphere and an unchanging superlunary sphere. As the planets belonged to the superlunary sphere their orbits had to be circular, because the circle was the most perfect geometric figure; it was regular and invariant (see [66, Part I]).

- The existence of closed time-like curves in Gödel-type universes. Whilst sections of the global world line always point in the forward direction of time, the total curve returns to its origin, where the Big Crunch is regarded as a time-reverse of the Big Bang. Local observers would only experience a passage of time but their inferences from local to global arrows could be misleading because in such a universe the experience of local arrows of time would be compatible with closed time-like curves (cf. [16, §2.6]). Such a Gödel universe would not be time-orientable because the simultaneity hyperplanes—Cauchy surfaces—would not be ordered sequentially, along a given temporal axis, defined by the average mass-energy distribution in an evolving universe.
- The existence of time-orientable space-time models. By contrast, as discussed below, a time-orientable universe is simply a model of space-time with a conventional time-direction, which does not tell us whether the actual temporal orientation of the physical universe corresponds to the model structure. According to Earman's 'Time Direction Heresy', the arrow of time becomes a geometric feature of space-time, making the appeal to entropy seemingly redundant [12]. This geometric approach makes space-time time-orientable by embedding the arrow of time in the modeled fabric of space-time.

As it turns out a temporally oriented space-time still requires 'energy fluxes', typically associated with the Second law, to establish temporal asymmetry. If a universe performs an evolution from a Big Bang to a Big Crunch its topology may either be that of a closed circle (in which case initial and final conditions are—nearly—identical and there may be no global arrow of time) or that of an open circle, if initial and final conditions are clearly distinguishable, for instance in terms of their entropic gradients but the universe collapses into a Big Crunch. Such a universe would display an arrow of time despite its contraction into black holes.

- The existence of different topologies of time. (There are 3 standard FLRW cosmological models, named after Friedman, Lematre, Robertson, Walker) If a universe evolves from a Big Bang to a Heat Death (or Big Chill) its evolution may be represented by an evolution curve [51, 52]. Such a universe has an unmistakable arrow of time because of the physical difference between initial and final conditions. Each FLRW model is equipped with a 1-parameter family of non-intersecting homogeneous space-like hypersurfaces, $\Gamma_t$, providing a simultaneity slice at $t$. According to FLRW models, the ultimate fate of the universe is dependent on the parameter $\Omega$, the ratio of actual to critical mass density. The critical mass density is a threshold value, at which the universe will still expand (see Fig. 5.1). The value of $\Omega$ can be smaller than 1 ($<1$), equal to 1 or greater than 1 ($>1$). If $\Omega$ is $<1$, then the critical mass density is greater than the actual mass density and the universe will expand forever. The question of expansion or contraction essentially depends on the amount of dark matter and dark energy in the universe. It cannot be directly observed, but its gravitational pull could be strong enough to bring the expansion of the universe to a halt and start the contraction phase. Current data indicate that the universe is actually accelerating so that at present the expansion seems set to

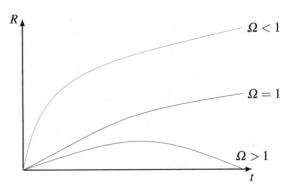

**Fig. 5.1** Critical values for $\Omega$ and the corresponding evolution of the universe where $R$ is its radius

continue forever. The value $\Omega = 1$ signifies that the universe will expand but at a more steady rate than at $\Omega < 1$. These values imply the total dissipation of energy, such that no energy differentials would be left to perform useful work and sustain life when the universe has reached equilibrium. The value of $\Omega > 1$ indicates that the actual mass density in the universe is greater than the critical mass density, which implies that the universe would re-collapse to what is sometimes called the Big Crunch. The parameter $t$ in FLRW models can go arbitrarily close to zero but its physical basis may break down in the early universe, as the Big Bang singularity is approached [56, §2.3]. The expansion of the universe will create local arrows of time.

## 5.2   Cosmic Arrows

As a Gödel-type universe shows it is possible for the local passage of time to exist in the absence of a cosmic arrow of time; but where there is a global arrow of time, there must be local arrows of time. In the latter case it is legitimate to infer, given appropriate criteria, the arrow of cosmic time from the anisotropy of local time. The existence of the cosmic arrow of time has often been associated with the increase in entropy but it is only one amongst a number of proposals.

- The oldest suggestion is to identify the cosmic arrow with an increase in entropy, which is often interpreted as an increase in disorder or randomness. The Austrian physicist Ludwig Boltzmann and the British astronomer Arthur Eddington both proposed that the increase in entropy could be used to identify the arrow of time. This proposal was an extension of the range of the Second Law from closed systems, like gas molecules in a sealed container, to the whole universe. Boltzmann characterized the Second law as proclaiming 'a steady degradation of energy until all tensions that might still perform work and all visible motion in the universe would have to cease' [4, p. 19] [7, §89]. He does not entertain any hope that the universe could be saved 'from this thermal death.' But in his reply to various objections to his theory, he cautions against assuming 'the unlimited validity of the irreversibility principle, when applied to the universe for an infinitely long time (...)' [5, p. 226]. In fact, Boltzmann held the whole universe to subsist in a state of equilibrium, with pockets of disequilibrium, which alone can sustain life. Under this assumption 'one can understand the validity of the second law and the heat death of each individual world without invoking a unidirectional change of the entire universe from a definite initial state to a final state' [6, p. 242]. However, three decades later A. Eddington believed that he had identified in the Second law of thermodynamics—the increase of entropy in isolated systems—an irreversible arrow of time [18, p. 68] [52, p. 697]. He considered that no natural processes would ever run counter to the Second law and that it held a 'supreme position among the laws of Nature' [18, p. 74]. But it is part of the evolution of Eddington's thinking about temporal arrows that his views underwent a change in perspective from, say, a Platonian to an Aristotelian view. That is, Eddington moved from an identification of the Second law with an increase in entropy to the employment of the Second law as a criterion or 'sign-post' for the arrows of time.

- Earman's time-direction heresy abandons the reliance on the Second law altogether [11, 12] [49, p. 719]. This is the proposal that a cosmic arrow of time is inherent in space-time in the sense that it is a geometric feature of space-time itself. Strictly speaking, the arrow of time is built into the model of space-time, which makes the space-time models time-orientable. This proposal has the advantage of being able to dispense with the controversial notion of entropy [14, p. 257]. But Earman's proposal also has severe disadvantages: Although this approach invests space-time with a conventional temporal direction, it does not invest it with a temporal arrow because a temporally orientable space-time is not the same as a temporally oriented space-time. The space-time model is time-orientable but this does not tell us what the actual time-orientation of the universe, which is being modelled, is. Standard FLRW models are time-orientable but their actual evolution depends on the value of spatial curvature, due to the ratio of actual over critical mass density ($\Omega > 1$; $\Omega = 1$; $\Omega < 1$). Proponents of this view accept that:

  (t)emporal orientability is merely a necessary condition for defining the global arrow of time, but it does not provide a physical, nonarbitrary criterion for distinguishing between the two directions of time [11, p. 2496] [1, 49].

Such non-arbitrary criteria may have to be derived from observable energy fluxes in the universe and other criteria.

- Thomas Gold proposed that cosmic expansion was to be the driver of the arrow of time. In Gold's view the cosmic arrow of time tracks the cosmological expansion [33–35]. Gold universe models are inspired by symmetry considerations. They usually postulate two-time boundary conditions, according to which the initial and final conditions of the universe are approximately identical. It is usually assumed that a Gold universe undergoes a cycle from a Big Bang to a Big Crunch, where each is marked by low entropy conditions (Fig. 5.2a, for symmetry reasons Gold universe models usually postulate a future condition, which then acts as a constraint on the current entropy curve). But a Gold universe could start and end in

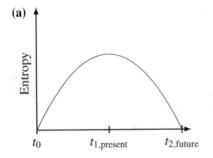 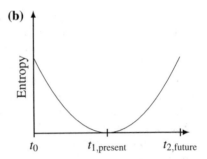

**Fig. 5.2** The Gold universe (**a**) and its inverse (**b**)

a high-entropy state, as long as its symmetric nature is preserved (Fig. 5.2b). The advantage of Gold universe models is that they satisfy a symmetry requirement as a default position so that they do not favour initial low-entropy over final high entropy conditions. If a low-entropy past in the Big Bang is postulated, without the constraint of a low-entropy final condition, the arrow of time looks like a consequence of the Second law of thermodynamics. However, if initial conditions are factored in, as St. Hawking complains, 'the second law is really a tautology.' Entropy increases with time because 'we define the direction of time to be that in which entropy increases' [41, p. 348].

However Gold universe models also have several disadvantages:

1. On the *empirical* side there is strong evidence of a low-entropy Big Bang but there is no evidence of a future low-entropy Big Crunch. As discussed below, pre-Big Bang scenarios in today's cosmology no longer stipulate a low-entropy Big Bang but derive it from pre-Big Bang conditions. There is also empirical evidence of an accelerated expansion of the universe, with the expectation that it will end in a high-entropy Big Chill.
2. On the *logical* side a Gold universe saves the symmetry of the boundary conditions by introducing an unexplained dynamic asymmetry in the evolution curve. The symmetry of the boundary conditions—they have either low or high entropy at the beginning and the end of the lifetime of the universe—requires a switch-over point, at which the evolution curve changes direction. But this approach provides no dynamic explanation why the evolution curve switches at $t_{1,present}$ from high to low entropy (Fig. 5.2a) or decreases from $t_0$ to $t_{1,present}$ (Fig. 5.2b). Both cases assume some form of retro-causality. In the first case, a low-entropy future constraint acts against the Second law of thermodynamics and returns the universe to a state of low entropy, where the Big Crunch is taken to be identical to the Big Bang. If it is assumed, as in the second case, that the Big Bang was a high-entropy state, then the current lower entropy-state would either be an unlikely fluctuation or the current state would have to act as an attractor state, which lowers the entropy from its high entropy beginning. Eventually the universe would resume its typical evolution towards a high entropy end state. Such

scenarios are imposed by the symmetry conditions of Gold universe models. If we take Gold universes to be symmetrical with respect to initial and final conditions, without specifying whether these are low or high entropy conditions, then such Gold universe models are committed to an unjustified explanatory asymmetry (despite the symmetry of boundary conditions) at a switch-over point ($t_{1,present}$ in both scenarios).

On the assumption of a low-entropy past the evolution of the universe displays behaviour in accordance with the Second Law but after a maximal expansion suddenly switches to anti-thermodynamic behaviour towards a low-entropy future; by contrast if the universe starts off in a high-entropy state its evolution will at first run counter to the Second law; this would require the presence, at its minimum, of an attractor state, which makes the universe behave in an anti-thermodynamic way, whilst its second phase towards a final high-entropy state will again be in accordance with the Second law. The unanswered question is what dynamics lie behind the switch-over in both scenarios.

- The Growing Block Universe. According to this proposal space-time is to be modelled as an Evolving Block Universe, which grows into an open future with the passage of time [23], cf. [56]. Hence the scenario $\Omega > 1$ is excluded. An initial low-entropy past (Big Bang) sets the conditions for a master clock, which provides the universal direction of time. The master clock has an effect on local arrows of time through the increase of entropy and other mechanisms. Space-time literally emerges as the master arrow evolves towards an uncertain future. The emergence of local arrows (passage of time) may be due to such mechanisms as wave function collapse, decoherence and increase in entropy. The postulation of a past hypothesis means that this proposal introduces an asymmetry from the start. Although there is empirical evidence for a low-entropy past of the universe, from a theoretical point of view the postulate of a low-entropy past (called the Past Hypothesis) is put in by hand. Theory, however, requires an explanation of a low-entropy beginning of the universe, as it is tentatively provided by pre-Big Bang scenarios.

Eddington's position partly overlaps with some of these proposals.

## 5.3 Eddington and the Arrow of Time

From this brief review it seems to be evident that the notion of entropy has not been banned from discussions of the arrow of time. It will be useful to re-examine it by paying close attention to Eddington's discussion of the anisotropy of time. As mentioned earlier both Boltzmann and Eddington had a tendency to identify the arrow of time with increasing entropy, which made the Second Law of Thermodynamics the only candidate for the arrow of time. But both also had reservations about this identification. Although Boltzmann characterized the Second Law as a 'steady gradation of energy' he did not think that the motion towards a 'heat death' applied to the whole universe [4, p. 19] [7, p. 89]. Rather he assumed that the whole universe existed in a state of equilibrium, with individual pockets of disequilibrium, where life may evolve.

Eddington's thinking, too, underwent a significant development. In his early essays on the theory of relativity he embraced the notion of a static block universe, which relegated the passage of time to a mental phenomenon, in the Kantian sense [65, pp. 138–147]. But in his later work he argued that in order to 'express the one-way property of time', an arrow of time must be added to the geometric representation of the four-dimensional world as a map without flow. It is in this connection that he assigns the Second Law a special place amongst the otherwise time-reversal invariant fundamental laws of physics. The increase in randomness or entropy will allow us to distinguish the past from the future. In *The Nature of the Physical World* Eddington grants the Second Law an apparently unassailable, exceptionless 'supreme position amongst the laws of Nature' [18, p. 74] and adds that as 'far as physics is concerned time's arrow is a property of entropy alone' [18, p. 80]. With the introduction of entropy into the physical world view a transition from a static to a dynamic universe had been made possible [18, p. 110]. Note, however, that the exclusive emphasis on entropy is not compatible with his epistemological views, according to which physical reality is an inferential construction from pointer readings and phenomena.

In his later book *New Pathways in Science* (1935) he introduces further refinements to these ideas. He acknowledges that the Second Law is a statistical law, but does not concede that it may have lost its supreme position amongst the laws of physics. The increase in entropy now becomes a 'signpost' for the arrows of time. Eddington explicitly moves away from his earlier view, which seems to suggest that time's arrow should be identified with the Second Law. At first it looks as if Eddington's change of mind was not brought about by the reversibility (Loschmidt) and recurrence (Zermelo) objections, which had forced Boltzmann to change the status of the Second Law from a deterministic to a statistical law. Eddington does not seem to agree with some of today's commentators who hold that the Second Law is not a fundamental law and that its statistical character makes it a poor candidate for an identification of the arrow of time. He continues to regard entropy as a 'unique local signpost of time' [19, pp. 68–71]. But at thermodynamic equilibrium–$dS/dt = 0$–'our signpost for time disappears and time ceases to go on'; nevertheless it continues to exist and like space, it extends since, for instance, atoms still vibrate [18, p. 79]. In such a universe observers could still tell the ticking of time but time would have lost its direction. Whilst entropy becomes a signpost for the local arrows of time, the expansion of the universe becomes a signpost for the cosmic arrow of time.

By admitting that the Second Law is a matter of probability, not certainty [19, p. 61] does Eddington's position not become vulnerable to Loschmidt's reversibility argument, which is based on the time-symmetry of the classical laws of physics? [67, pp. 180–181]

On closer reading we find that Eddington does offer an argument against reversibility, in terms of phase space considerations, which anticipates a similar argument later proposed by Landsberg [45] cf. [52, p. 701] [10, pp. 336–342]. It turns out to be an alternative argument for the cosmic arrow of time, which makes no explicit reference to entropy but rather relies on the expansion of the universe. This argument is interesting because it no longer refers to increasing disorder or randomness; and it can be related to today's phase space and typicality approaches to the arrow of time. What does this 'phase space argument' say?

First, Eddington rejects Boltzmann's argument that our habitable universe is a giant entropy-defying fluctuation from a universe in thermodynamic equilibrium. Eddington does not accept that the Second Law would allow fluke fluctuations. He confines chance fluctuations to the beginning of the universe [19, p. 66]. It is in this context that he introduces a new argument against the unnamed reversibility objection, namely the expanding universe, which he understands in terms of volume changes [19, Chap. III, §5]. He regards both the 'dissipation of energy' and the 'expansion of the universe' as irreversible processes, and hence as signposts for the arrow of time. The expansion of the universe is a global signpost, whilst the increase in entropy is a 'unique local signpost'.[2] But how do these signposts avoid the reversibility objection? Eddington argues that, if an expanding universe is taken into consideration, we are no longer forced to conclude 'that every possible configuration of atoms must repeat itself at some distant date'.

> In an expanding space any particular congruence becomes more and more improbable. The expansion of the universe creates new possibilities of distribution faster than the atoms can work through them, and there is no longer any likelihood of a particular distribution being repeated. If we continue shuffling a pack of cards we are bound sometime to bring them into their standard order but not if the conditions are that every morning one more card is added to the pack [19, p. 68].

A scenario which is theoretically possible, according to Loschmidt's reversibility objection, does not thereby become typical, statistically relevant behaviour in a dynamic universe. Eddington seems to hold that as a system expands into the accessible phase space, with the expansion of the universe, the probability of a return to initial conditions becomes less likely. These considerations are noteworthy for two reasons: (a) they indicate a new proposal for the explanation of the cosmic arrow of time; (b) they link Eddington's refined views to current considerations of the arrow of time in terms of phase space volumes and typicality arguments. Eddington's argument is plausible from a practical point of view, but the question remains how the expansion of the universe circumvents Loschmidt's fundamental point of the temporal blindness of probability relations, irrespective of the occupied phase space volume. Does Eddington need some stronger argument to avoid Loschmidt's objection? This question will be investigated in the remainder of this chapter. Note that the argument in terms of phase space is preferable to the language of entropy. Frequently the notion of entropy is associated with an increase in disorder or randomness when in fact an increase in order is perfectly compatible

---

[2]The expansion of the universe is however related to phase space arguments, in today's language, which provide an alternative reading of the Second Law. Eddington also distinguishes physical from phenomenal time. Physical time is inferred from physical criteria, like entropy and the expansion of the universe. Phenomenal time is the subjective impression, located in the mind, of the 'flow' of time [18, Chap. V] [19, Chap. III, §] [60]. Later Eddington suggests that the 'feeling of passage' may be explained by entropy changes in the brain [18, p. 100] [20, Chap. XII].

with an increase in entropy. The universe is said to have started in a low-entropy Big Bang but although it has increased its entropy (it occupies a larger volume of its phase space) it also has increased its manifest order, as evidenced by the order of galaxies. Eddington suspects that there is some undiscovered relation between the two criteria: the increase in phase space volumes and the increase in entropy [19, p. 68]. This relation, we may propose, is provided by implications of Liouville's theorem and typicality approaches to the arrow of time. Phase space arguments need to be enhanced by typicality arguments.

## 5.4  Phase Space and Typicality

In his attempt to formulate a solution to the arrow of time, Peter Landsberg refers to the well-known illustration of an expanding gas in a sealed contained and suggests that a 'basic principle is at work here':

> Suppose a system develops without interference from outside. Then it chooses among its available equilibrium states in proportion to their realisabilities. I shall call this principle **P**, and we note that it makes at least no explicit reference to time. But it implies that if one available state is enormously more realizable than any other, then we do not go far wrong by predicting that this state will be *the* equilibrium state.

He goes on to suggest that this is the state, which has the greatest realisability.

> In this way, then, the principle extracts a direction of time even though the molecular collisions which give rise to the diffusion of the gas are each time-reversible. It does so by statistical averaging in proportion to the realisabilities [44, p. 75, bold and italics in originals].

Entropy should be associated with realisability and phase space volume rather than disorder. As will be shown below, realisability implies that it is important to distinguish between accessed and accessible phase space, a distinction, which can be applied to cosmology [51, p. 408ff]. In order to mark the difference between what is theoretically possible and what is statistically probable Landsberg defines a notion of:

> 'weak' T-invariance of a complex process by the requirement that its time inverse (although perhaps improbable) does not violate the laws of the most elementary processes in terms of which it is understood. Thus, diffusion and heat conduction, while not T-invariant, are at least weakly T-invariant [44, p. 8] [50, p. 402].

This proposal can be related to Liouville's theorem, to which both Eddington and Landsberg implicitly refer. It stipulates the volume invariance but denies the shape-invariance of a classical dynamical system (see Fig. 5.3).

Liouville's theorem in classical mechanics states that a volume element along a flowline conserves the classical distribution function $f(r, v)drdv$ [42, p. 408], [2, p. 73f]:

$$f(t + dt, r + dr, v + dv) = f(t, r, v) \tag{5.1}$$

Liouville's theorem says that the dynamic evolution of a classical system preserves the *volume* of the initial phase-space region but not its *shape*. The shape of the phase-space region can become very unstructured, disordered, a physical state for the description of which sometimes the term 'fibrillation' is used [60, pp. 55–56] [53, pp. 25–34]. A rather uniform (smooth) region in the initial stage of dynamic evolution can become very fibrillated since the shape of phase space regions is not preserved. The division of phase space into different cells—with different shapes—is known as 'coarse-graining'. That is, the volume of the available phase space regions is invariant over time even though the motion of a point along an evolution curve within this volume can start from smooth initial states or end up in fibrillated final configurations.

In other words, if we consider trajectories in phase space, which include both position and momentum of particles, then the equation of motion of such systems can be expressed in terms of its Hamiltonian, $H$. $H$ expresses the conservation of total energy of the system. Liouville's theorem then states that the volume of the phase space, which an ensemble of trajectories occupies, remains constant over time (see Fig. 5.3). Liouville's theorem shows that the volume of the phase space regions is invariant over time even though the evolution of the trajectories within this volume can start from different initial states or end in different final states. But an immediate consequence of this theorem is that even though the volume is preserved the shape of this phase space region is not preserved and this implies a dynamic evolution of the trajectories within this region. Past and future are not mirror images of each other [14, §4]. For two shapes cannot differ from each other without an evolution of the trajectories. It also implies that a reversed evolution of the trajectories will preserve the volume but not the shape and hence that reversed trajectories need not be invariant with respect to the shape of the phase space region.

Even though the talk here is of spreading and fibrillation the question remains in which way this approach can evade the reversibility objection. When phase space

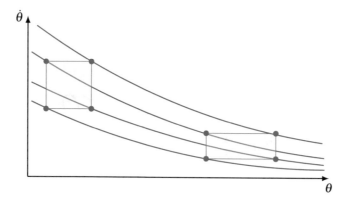

**Fig. 5.3**  An illustration of Liouville's theorem: volume invariance but not shape invariance. *Source* [61, p. 206]

arguments are combined with typicality considerations, the view emerges that all evolutions may be equally possible in theory but they are not equally probable in practice.

### 5.4.1  Kinematics I: Phase Space Arguments

All attempts to *identify* the arrow of time with a particular physical process have led astray.

- Plato identified the arrow of time with the circular motion of the planets. But it turned out that the planets do not move in circular orbits and that their elliptical motion around the sun is not uniform.
- The attempt to identify the arrow of time with the Second Law has led to a number of objections:

  - The Second Law is not a fundamental law; 'all' fundamental laws are time-reversal invariant and make no distinction between past and future.
  - The Second Law is a statistical law; whilst on empirical grounds entropy-decreasing fluctuations may be neglected, they cannot be excluded on logical grounds so that Gold universes remain a logical possibility.
  - On logical grounds also probabilities are blind to temporal directions, which implies that time-reversed evolutions must be considered (Loschmidt's Demon—see Fig. 5.4).

Thus Plato's strategy to *identify* the passage of time with some physical process is mistaken. But Aristotle's strategy to mark some physical process as a criterion for the

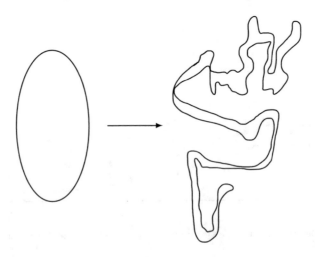

**Fig. 5.4**  Can a Loschmidt Demon return the fibrillated final state to the ordered initial state?

passage of time remains a valid option. As we have seen, Eddington began to think of entropy as a good criterion for the arrow of time. But Eddington was mistaken in taking entropy as the sole criterion of the anisotropy of time; it is not in line with his 'operationalist' philosophy. It is a good criterion FAPP (for all practical purposes), since it is weakly $T$-invariant. Thinking in terms of criteria for the arrows of time means that several physical criteria can be taken into account; this is the path, which Eddington eventually took.

Practically, irreversible behaviour is typical. Boltzmann estimated the amount of time it would take for a volume of gas, containing $10^{18}$ molecules, to return to its initial state (position and momentum variables). Boltzmann considered that each molecule (with an average molecular velocity of $5 \times 10^{14}$ cm/s in both directions) would return to within $10^{-7}$ cm of each initial position variable and within $10^2$ cm/s of each velocity variable. The estimated time for a return to such a configuration would require $10^{10^{19}}$ years, which is well beyond the estimated age of the universe ($\sim 10^{10}$ years) [57, pp. 52–53].

Let us first consider the topology of phase space (a) and secondly typicality arguments (b) in order to see how the $T$-invariance of physical laws is compatible with asymmetric solutions, and how the irreversibility of physical processes can be explained without the traditional understanding of entropy as increase in disorder.

A classical system in mechanics is standardly described in terms of a $6N$-dimensional phase space (a), $\Gamma$, in which each individual particle has three position coordinates $(x, y, z)$ and three momentum coordinates (since 'momentum = mass $\times$ velocity' is a vector quantity). Single particle systems or many particle systems are represented by a single point $\chi$, its micro-state, which moves around in phase space according to the deterministic laws of Hamiltonian mechanics. The phase space usually comes endowed with a Lebesgue measure, $\mu$, which roughly is a volume measure of the phase space, available to the systems. For Hamiltonian systems the Lebesgue measure is invariant under the dynamics; this statement is equivalent to Liouville's theorem. When $\mu$ is restricted to a normalized measure $\mu_E$ on a $(6N - 1)$-dimensional hypersurface $\Gamma_E$—to which the motion of the system is confined—$\mu_E$ is also invariant under the dynamics. When speaking of the phase space of a particular system, some care must be taken as to the precise meaning of this term. When Maxwell's Demon sits in a sealed container, with a small opening in a partition wall, which allows the Demon to separate the slow from the fast molecules, he is confined to a three-dimensional Euclidean coordinate space. When the partition is removed (together with the Demon) the gas molecules will spread through the whole available three-dimensional volume. Before the removal of the partition, the phase space, which the molecules occupied, was smaller than the total phase space available to them. This three-dimensional example makes clear that a distinction should be drawn between the phase space *occupied* and the phase space *available* to the molecules. In a more mathematical sense, phase space, $\Gamma$, is conceived as a $6N$-dimensional abstract space, which describes all the particles, which belong to the system. If energy conservation is taken into account, 'the motion of the system is confined to a $(6N - 1)$-dimensional energy hypersurface $\Gamma_E$ which describes the phase space available to the evolution curves (or phase flow $\Phi_t$)

[68, p. 471]. When phase space (or the space of states) is conceived as the phase space of all the possible states, which a system could hypothetically occupy, the reversibility of the fundamental equations requires that this available phase space remain invariant [10, pp. 336–342] [52, p. 701] [51, p. 407ff]. The phase space must remain invariant if the trajectories are allowed, in theory, to return to their initial conditions. According to the Liouville theorem, $\mu$ is a volume-preserving measure. But asymmetry obviously implies some evolution: what changes is not the number of *possible* configurations but the number of *actual* configurations. That is, the volume of phase space, which the configurations actually occupy before reaching equilibrium, is smaller than the volume of phase space available to them. Given the expansion of the universe the phase space available to cosmological systems is obviously much larger than the volume of phase space currently occupied. Due to the difference between phase space occupied and phase space available, the system evolves, such that the equilibrium macro-state is larger than any other state. As a system, like the universe, undergoes expansion, it begins to occupy different volumes, due to the different configurations it can occupy—for instance $|\Gamma_{M_{equilibrium}}| \gg |\Gamma_{M_{initial}}|$—and these different volumes allow the construction of volume ratios—$|\Gamma_{M_i}|/|\Gamma_{M_{eq}}|$—which become important for asymmetry arguments. Eddington regarded these volume differences as another criterion—apart from the traditional understanding of entropy—for a cosmological arrow of time [19, pp. 67–68].

The phase space argument must be applicable to the universe as a whole, which requires that the universe be treated as a closed system. This means that the 'degrees of freedom that are available to the whole universe are described by the total phase space' [52, p. 701] [51, pp. 407–411, pp. 466–467]. However we face the same problem as with phase space in a classical system. The phase space of the universe remains invariant means that the trajectories are allowed in theory to return to their initial conditions. This is Loschmidt's reversibility objection. As seen from the current state the entropy is expected to increase into the future, leading to a high entropy-state, like the Big Chill. But in a time-reverse direction we should expect entropy to be high in the past, which contradicts the evidence and raises the question of how the universe could move from high entropy in the past to lower entropy now. Therefore a constraint has to be imposed on the initial condition—the Past Hypothesis—which states that the entropy was low in the past. The objection to this imposition is called the 'double standard' fallacy. Initial conditions are favoured over final conditions, which run counter to a fundamental symmetry postulate, as in Gold universe models [54].

Although notions like 'fibrillation' and 'spreading' have an intuitive physical appeal, in the statistical-mechanical literature spreading usually refers to the realisability of the macro-state with respect to the available combinations of micro-states. Realisability describes the number of micro-states, which are compatible with or make up a given macro-state, as reflected in Boltzmann's definition of entropy: $S = k_B \log V$ (where $V$ is the volume of the coarse-grained region, which contains X). It is tempting to argue that the realisability of different configurations 'extracts a direction of time even though the molecular collisions which give rise to the diffusion of the gas are each time-reversible' [45, p. 75] [43]. For if a system has a greater degree of realisability available to its macro-states—a greater amount of spreading into the available phase space—this evolution could serve as a criterion for an arrow of time. However, as it stands, the phase space argument is vulnerable to reversibility objections. The realisability or phase-space argument must assume that there is no future constraint, which acts on the current state. The realisability is greater towards the future than towards the past but from a statistical point of view realisability should be equally likely in both directions. The question arises whether realisability can be saved if realisability in one direction can be said to be more typical than realisability in the opposite direction. Phase space arguments require the postulation of asymmetric boundary conditions but here typicality arguments come to play their part. They show that phase space arguments, supported by typicality, provide more plausible accounts of asymmetry than the symmetry required by Gold universe models.

## 5.4.2 Kinematics II: Typicality Arguments

Typicality arguments (**b**) are concerned with the overwhelming majority of cases in a specified set. This implies the ratio of overwhelmingly many to a small number of divergent cases, hence a large number ratios, in particular volume ratios. According to one formulation of the typicality view [36] 'for any [micro-region] $[\Gamma_{M_i}]$ the relative volume of the set of micro-states $[x]$ in $\Gamma_{M_i}$ for which the Second Law is violated (...) goes to zero rapidly (exponentially) in the number of atoms and molecules in the system' (quoted in [26, p. 84]). Such a notion of typicality resembles the notion of weak $T$-invariance [44, p. 8], since a weakly $T$-invariant process, such as diffusion and heat conduction, allows for its (improbable) time-reversal without violating the 'laws of elementary processes'. But such processes are never observed in nature, hence their realisability is practically zero or would require practically infinite amounts of time, as Boltzmann's estimate illustrates. These characterizations, which focus on volume ratios, seem to appeal to a notion of $volume - ratio$ typicality, since they make an implicit appeal to the ratio of $occupied$ and $available$ phase space, and the non-invariance of the structural shape of initial and final conditions. They may hold the key to the avoidance of the double-standard fallacy—the unjustified favouring of initial over final conditions—without assuming a symmetric Gold universe.

Recall that Eddington treated the volume differences $|\Gamma_{M_i}|/|\Gamma_{M_{eq}}|$ in an expanding universe as a separate criterion for the cosmic arrow of time. The realisability argument appeals to 'molecular collisions' to explain the evolution towards equilibrium. In the language of typicality, the realisability of more macro-states from fewer macro-states is more typical than the realisability of fewer macro-states from more macro-states because of the difference between occupied and available phase space. The increasing arrow of time can therefore be understood as a function of the ratio of occupied to available phase space. These occupied regions of phase space are regarded as statistically irreversible, whilst the equations of motion, which govern the trajectories, remain time-reversal invariant [50, p. 408]. But these time-reversible equations of motion are compatible with the statistical irreversibility of the macro-systems. In terms of the temporal blindness of probabilities this should matter only pragmatically—in terms of the time needed—but not on logical grounds. It seems that as long as the argument is formulated in terms of probabilities, a postulation of symmetry between initial and final conditions during temporal evolutions is unavoidable. Hence the appeal to typicality.

One advantage of volume-ratio typicality arguments[3] is that they are not based on probabilities; hence the problem of the $T$-invariance of probabilities does not immediately arise. Typicality approaches do not start from the assumption of the temporal neutrality between initial and final conditions. They do not assume that all evolutions are equally probable. Earlier it was suggested that the topology of phase space should be characterized in terms of Liouville's theorem. Usually Liouville's theorem expresses the fact that the evolution of the state of a system is volume-preserving but Liouville's theorem has an important corollary—the phase space is not shape-invariant. Hence the topology of the system changes over its thermodynamic evolution. According to typicality approaches the evolution to a fibrillated state is typical, whilst a return to a smooth, ordered state is atypical. Hence typicality approaches do not give rise to a demand for parity-of-reasoning: a quasi-identity of initial and final conditions is highly atypical, although it is not excluded. The number of degrees of freedom in a fibrillated state is much greater than in a smooth state, and hence their return to a more ordered state requires 'perfect aiming', which, given the physical constraints on a system, make it extremely atypical. It would require a Loschmidt Demon, whose work in an expanding universe, according to Eddington, becomes very difficult, since Liouville's theorem alone does not guarantee his success. A proponent of a $T$-symmetric model will ask by what right we can assume that evolution towards the future will be typical. On the empirical level, a proponent of typicality can point to the cosmological fact that the universe is expanding at an accelerated pace and will, in the fullness of time, reach what was dubbed the 'heat death' in the 19th Century. On an epistemological level the proponent of typicality will, as will now be argued, appeal to plausibility arguments.

Do typicality arguments establish a cosmic arrow of time? Typicality arguments have a negative and a positive aspect. On the negative side they provide arguments against the symmetry of the Gold universe, which by assumption has (nearly) identical boundary conditions, where these boundary conditions can either be in a state of high entropy or low entropy. Consider typicality in a Gold universe.

- If we start from the assumption, by the argument from parity, of high entropy states both at the beginning and the end of a system's evolution, then on Boltzmann's assumption the current state of a system, like the universe, is in a lower state of entropy, hence in a more ordered state. Such fluctuations have been

---

[3]Generally speaking, a set is typical if it contains an 'overwhelming majority' of points in some specified sense. In classical statistical mechanics there is a 'natural' sense: namely sets of full phase-space volume. That is, one considers the Lebesgue-measure on phase-space, which is invariant (by Liouville's theorem) and which, when 'cut to the energy surface', can be normalized to a probability measure; then sets of volume close to one are considered typical' [63, p. 803]. Even if typicality claims are formally called 'probability measures', this does not make them probabilities in the philosophical sense [68, p. 475Fn].

dismissed as implausible [10, p.221–224] [21, pp. 115–116] [47, p. 142] [54, p. 35]. A Boltzmann fluctuation would require an evolution of a more fibrillated state to a less fibrillated state or, from the point of view of typicality, a change in the ratio of phase space volumes. A Boltzmann fluctuation would mean an atypical switch in the ratio of phase space volume at some point on the evolution curve, which would itself become atypical (Figs. 5.2a, b).

- This asymmetry in typical evolutions remains if we postulate an alternative Gold universe, with equal initial and final conditions of low entropy. Now the current state of the system is in a higher state of entropy than the initial condition, which is typical behaviour but the system is supposed to return to a lower state of entropy, hence an atypical, anti-thermodynamic evolution. Such behaviour requires the postulation of constraints in the distant future, attractor states, which by retro-causality would have an effect on the current state [58]. There is no evidence for such future entropy constraints. Whether or not evidence for such a future constraint on present conditions exists, another atypical switch in volume ratios occurs.

These topological considerations do not favour initial over final conditions; they do not require 'perfect aiming' or an appeal to a Loschmidt Demon, since they focus on the volumes of phase space occupied by respective regions and consider the volume ratios in a Gold universe. A Gold universe of either type must assume an atypical switch in volume ratios. This reversal of volume ratios means that a proponent of a Gold universe faces explanatory anomalies. Thus Gold universe models face a switch-over problem: what causes its occurrence?

On the positive side typicality arguments appear to be kinematic descriptions of volume ratios and provide plausibility arguments for the existence of an arrow of time. In the absence of a future constraint (low-entropy condition) coupled with evidence for the Big Bang, an entropy gradient exists between the early entropy ($10^{88}$) and the final entropy ($10^{120}$) of the universe [10, p. 63] [51, Chap. 7]); there are also observable 'inhomogeneities and anisotropies in the later time universe' [3, §1]. The criteria provide strong reasons to infer that there is an arrow of time. But such kinematic descriptions do not provide dynamic reasons for their existence.

## 5.5 Dynamic Considerations

While[4] phase space and typicality arguments provide plausible kinematic accounts of the existence of a cosmic arrow of time, their plausibility needs to be support by dynamic considerations. A number of such dynamic accounts have been made but they return us to the earlier distinction between local and global arrows of time. It is worth remembering that the emergence of arrows of time is still an unsolved question and some of the proposals are at a speculative stage.

---

[4]Some material in this section is based on [67, Chap. IV].

### 5.5.1 Local Arrows

Recall that local arrows of time must be distinguished from the cosmic arrow of time. A cosmic arrow cannot simply be inferred from local arrows of time, since the passage of time is compatible with a Gödel-type universe models. They display time-like closed curves and lack a global arrow. But we do not live in a Gödel-type universe. The question therefore arises whether there is a master arrow of time, from which all other arrows could be derived. The parallelism of the many arrows of time—their uni-directional nature—raises the question of whether they have a common origin or whether one arrow is more fundamental than another. This section will consider some dynamic proposals to ground local and global arrows of time; it will defend the notion of multi-fingered time, and hence deny that there exists a master arrow of time.

Eddington regarded entropy as a 'signpost' for local arrows of time, whilst the expansion of the universe becomes an indicator of the cosmic arrow. Local arrows of time manifest themselves in various ways: the psychological (or phenomenal) arrow expresses a person's subjective feeling of the unidirectional 'flow' of time from past to future. It can vary from person to person, depending on their psychological or physiological state. The historical arrow of time indicates biological, geological and social evolutions in the environment. Yet, irrespective of our impressions of the 'flow' of time, we are also surrounded by many irreversible physical processes. Such irreversible processes—most people are born young and die old; hot liquids grow cold; food rots and buildings dilapidate—constitute an objective physical arrow of time. These irreversible processes seem to describe a transition from more to less orderly states, since they are based on the thermodynamic notion of entropy.

If we return to the gas container as an illustration of spreading in phase space (or coordinate space) an intuitive reason for the spreading of the gas molecules into the available coordinate space is due to the collisions between the particles: 'by far the main contribution to the entropy comes from the random particle motions', which make a return to the original distribution atypical [18, p. 68] [43, 45] [51, p. 402] [60, p. 174]. The random collisions change their direction isotropically and hence they will be sent along evolution curves, in which they are not constrained by either the walls of the container or other molecules—and this is the available and accessible phase space. But as the volume of the final phase space region is overwhelmingly larger than that of the initial region, the probability of a return of the system to a smaller region is practically zero. Loschmidt's Demon will nevertheless object that initial conditions have been prepared in the gas container, which turns the system into an asymmetric system. Probabilities, the Demon will continue, are blind to temporal directions; hence the molecular collisions should be as likely to return the system to a lower state of entropy—towards a smaller region of phase space—as they are to propel it to a higher state of entropy. The Demon is powerless, in practice, against the non-invariance of the shape of a region of phase space.

On analogy with Eddington's argument, in an expanding container a return of the molecules to their original position would equally become increasingly less likely. As the shape of the macro-region is not preserved, we expect a dynamic evolution of the states along evolution curves within the accessible phase space region. This will result in a ratio $|\Gamma_{M_i}|/|\Gamma_{M_{eq}}|$. But taking the ratio of volumes of phase space, in itself, does not explain why thermodynamic systems tend to obey the Second Law. Two possible explanations are: (a) $\varepsilon$-ergodicity and (b) environmental decoherence.

The dynamic evolution of such systems, governed by Hamiltonian dynamics, must also be typical. According to one proposal, Hamiltonians are typical if they are $\varepsilon$-ergodic **(a)**, that is 'ergodic on the phase space except for a small region of measure $\varepsilon$' [26, p. 628] [28, p. 922]. The original ergodic hypothesis, which is provably false, was eventually replaced by a quasi-ergodic hypothesis, which holds that a trajectory in phase space, $\Gamma$, starting from a point P will eventually come arbitrarily close to a point G, which represents any other point in phase space [60, p. 77]. The ergodicity of dynamic systems has two consequences:

1. 'if a system is ergodic, then for almost all trajectories, the fraction of time a trajectory spends in a region $R$ equals the fraction of the area of $X$ that is occupied by $R$.
2. 'almost all trajectories (i.e. trajectories through almost all initial conditions) come arbitrarily close to any point on the energy hypersurface infinitely many times (...)' [24, pp. 121–122].

A modern version of the ergodic hypothesis refers to a spreading function in phase space:

> a system is $\varepsilon$-ergodic if its phase space has an invariant subset of measure $1-\varepsilon$ such that, given almost any point in that subset, the trajectory that goes through that point sooner or later crosses every region of the subset that has positive measure [64, p. 2] [27, p. 634 for a more technical definition].

The intuition behind $\varepsilon$-ergodicity is to capture the idea, expressed in weak $T$-invariance, that deviations from typical behaviour are allowed around a small value $\varepsilon \geq 0$ [24] [28, p. 923]. A Hamiltonian displays typical behaviour if it is $\varepsilon$-ergodic [28, p. 927]. Although it is typical for a thermodynamic system to be ergodic in this sense, there is a small set, which does not have an ergodic solution. In other words, there is a spreading function [48, p. 1750], which uses the distinction between available and occupied phase space, micro-states and macro-states. But this spreading function must not be considered as a monotonic increase in entropy but as a typical solution, which in the case of thermodynamic systems, is weakly $T$-invariant (or $\varepsilon$-ergodic). If Hamiltonian evolutions are $\varepsilon$-ergodic, they constitute one criterion for the local passage of time.

The notion of decoherence **(b)** was introduced in the 1970s as a general mechanism, whereby macroscopic phenomena emerge from the interaction of quantum systems with their surrounding environment over relatively short decoherence time scales. Decoherence stands for the loss of interference terms—the dissipation of coherence—which is characteristic of the transition from quantum to classical systems. Decoherence in the environmental sense is understood as an interaction of a quantum system with its environment. It is illustrated in many versions of the two-slit experiment whereby the characteristic interference terms disappear at the classical level so that the familiar macro-phenomena are observed. The likelihood of measuring the superposed states in Schrödinger's famous cat paradox is close to zero. It is appropriate to see in decoherence an indicator of the anisotropy of time since it involves 'branching events' or the emergence of consistent histories in quantum mechanics. Environmental decoherence may explain the emergence of local arrows of time.

The decoherence mechanism is thus an attempt to explain the emergence of classical observable states from the underlying superpositions of quantum systems. The potential states of a system are expressed in the state vector $|\Psi\rangle$ or the wave function, $\Psi$. These superpositions are illustrated most dramatically in Schrödinger's cat paradox. As long as the box remains sealed, the cat's initial quantum state is expressed as a superposition of two alternative states [72, p. 58, cf. p.29]:

$$|\psi|cat_{initial-state}\rangle \rightarrow \alpha|atom_{not-decayed}\rangle \, |cat_{alive}\rangle + \beta \, |atom_{decayed}\rangle |cat_{dead}\rangle .$$
(5.2)

As quantum systems are often entangled, quantum entanglement is the 'basic mechanism underlying decoherence' [59, pp. 13, 28]. Quantum entanglement means that two subsystems, $S_1, S_2$, which constitute a composite system, cannot be treated as two individual systems. According to the Projection Postulate, a quantum measurement 'forces the system into one of the eigenstates of the measured observable (...)' [72, pp. 182–183] [37]. Decoherence, as a form of measurement, induces 'macroscopic systems into mixtures of states—approximately eigenstates of the same set of effectively classical observables' [72, p. 184]. Decoherence means that:

> the environment acts, in effect, as an observer continuously monitoring certain preferred observables [the classical pointer basis], which are selected mainly by the system-environment interaction Hamiltonian [72, p. 184] [32, 40].

Moreover the 'measurement' carried out by the environment invalidates data from exotic superpositions of states 'on a decoherence time scale'. The decoherence time scale, $\tau_D$, is very short, typically of the order of $10^{-20}$ s for a small particle at room temperature. This means that spread-out 'superpositions of localized wave packets will be rapidly destroyed as they will quickly evolve to mixtures of localized states' [59, p. 138] [72, p. 187].

As mentioned above the thermodynamic-like behaviour of local systems and environmental decoherence do not imply statements about any global arrow of time because of the theoretical possibility of close time-like curves. Furthermore, the universe is a unique system so that it seems that the notion of typicality is inapplicable. However the universe can always be defined as a closed thermodynamic system, in which case it is one amongst many thermodynamic systems, which display typical and atypical behaviour. It is therefore legitimate to ask whether a global arrow of time exists.

## 5.5.2   The Global Arrow of Time

Phase space arguments often assume initial conditions, in which thermodynamic systems are prepared. From a symmetric point of view this need poses a problem for a consideration of a global arrow of time since, it seems, that a global arrow of time only gets off the ground on the assumption of a Past Hypothesis (or the stipulation of low-entropy initial conditions). As Hawking complained the postulation of a low-entropy Big Bang, with its singularity, turns the Second law into a tautology because once such a low-entropy initial condition exists the Second law is guaranteed to take it to higher irreversible levels of entropy. The severity of this objection has, however, subsided because of recent attempts to explain the emergence of low entropy conditions in the early universe. Two such attempts are (a) Penrose's Weyl curvature hypothesis and (b) the notion of cosmological decoherence.

Penrose's gravitational clumping argument **(a)** explains the increase in entropy from smooth initial conditions at least in *our* universe. Penrose postulates a Weyl curvature tensor—essentially describing tidal distortions of space-time—which is set to zero for initial small volume of phase space at the Big Bang and this would account for the small entropy value at the Big Bang. Gravitational clumping then explains the increase in entropy: the original material clumps into stars and galaxies with the final collapse into black holes; the latter provide the highest source of entropy in the universe.

Within the framework of the General theory (GTR), the Weyl curvature hypothesis introduces a basic asymmetry and therefore justifies an arrow of time [50, p. 54ff] [52, §28.8, cf. §17.5, §19.7]. The Weyl tensor is that part of the Riemann tensor which describes gravitational degrees of freedom that occur even in the absence of matter. Space-time curvature can arise either in the presence of matter (described by the Ricci tensor) or in the absence of matter by gravitational tidal distortion (Weyl tensor). The Weyl tensor describes geodesic deviations of world lines, such as light rays (null geodesic) as may be caused by the gravitational fields. The Weyl tensor describes the tidal distortions on curvature, especially as they would appear at the beginning (Big Bang) and the end (Big Chill) of the universe. Penrose's hypothesis is that at the Big Bang, when the initial conditions of the universe are very smooth, the Weyl curvature must be zero or close to zero. But as the universe evolves— from the initial uniformity of matter at the Big Bang to the black hole singularities at the end of its life—there is an increasing gravitational clumping of matter. As more gravitational clumping takes place, the amount of Weyl curvature increases, and eventually diverges to infinity, 'as the black hole singularities' of the Big Chill are reached. Furthermore the Big Bang singularity is marked by a state of very low entropy, whilst the generic black hole singularities of the Big Chill are characterized by very high entropy (see Fig. 5.5).

**Fig. 5.5** A simplified representation of a universe with a neat Big Bang, with low entropy and zero Weyl curvature, and a messy Big Chill ending in black hole singularities, with high entropy and Weyl curvature. Adapted from Penrose [52, p. 729]

The universe starts at a small, uniform, hence low-entropy point in time and evolves towards a much larger high entropy region. During this evolution the Weyl curvature increases (clumping of gravitational matter) and reaches its maximum entropy in the formation of black hole singularities at the Big Chill [52, §27.7].

According to this hypothesis the Weyl curvature of the actual universe would be close to zero at the Big Bang but would diverge wildly at the Big Chill. 'As a result there would be a Second Law of thermodynamics, and it would take the form that we observe' [52, p. 768]. If this hypothesis were correct, the universe would be clearly time-asymmetric since its beginning could clearly be distinguished from its end by the divergence of the Weyl curvature. There would be a striking difference between the beginning and the final state of the universe. The universe would display an arrow of time. According to Penrose, the Second law defines 'a statistical time-directionality', which requires very special, low-entropy conditions at the Big Bang [52, pp. 730, 757, 777].

Penrose therefore sees a close connection between the thermodynamic and the cosmological arrows of time, since the evolution of the universe tracks the Second law of thermodynamics. But Penrose's Weyl curvature hypothesis has met with some critical reservations. Firstly, Penrose requires the Weyl curvature tensor to act as a constraint on the initial condition of the universe, from which the Second Law of Thermodynamics would follow. The Second Law would be an asymmetric law, quite different from the time-symmetric fundamental laws, whose existence is assumed in physics [50, p. 55]. A postulation of different kinds of laws operating in the physical universe flies in the face of the unification process of modern science. The status of the Second Law has also been subject to considerable debate. Its status as a fundamental law has been questioned [15, 54, 55, 62]. Secondly, it is ultimately unsatisfactory to postulate a low-entropy beginning of the universe. Thus Penrose introduces a lawlike asymmetry into the picture in stark contrast to the symmetry demands, associated with the Gold universe. However the search for the origin of the Big Bang in cosmology has recently led to a number of pre-Big Bang scenarios, which all aim to avoid the stipulation of a singularity. Penrose himself has introduced an oscillating universe; others have proposed baby universes. Some of these scenarios have two essential features. The universe is no longer born in a singularity but the Big Bang emerges from an earlier universe. These scenarios envisage global arrows of time within particular universes. Some of them restore symmetry across the multiverse. Talk of baby universes, the multiverse, cyclic and oscillating universes are decidedly at a speculative state.

Quantum cosmology is concerned with the boundary conditions of the universe, either at the very beginning or at the very end of its evolution, as well as with the emergence of the 'classical' realm.[5] When the initial and final conditions of the universe differ in their physical characteristics, an asymmetric evolution of the universe can be expected, creating an arrow of time. How can asymmetric boundary conditions arise? They would arise, if Penrose's Weyl curvature hypothesis were a correct description of these conditions. But the boundary conditions of the universe are so extreme that they fall outside of the realm of General relativity and are best considered in the context of quantum cosmology. A considerable consensus seems to prevail that the arrow of time arises from the existence of asymmetric boundary conditions of the universe, in conjunction with some regularity. Candidates for such a dynamic regularity are the expansion of the universe or the Second law of thermodynamics. Both are seen as problematic with respect to the arrow of time.[6] But if the arrow of time is to emerge from some fundamental quantum state of the universe, cosmological decoherence may provide the appropriate mechanism for its appearance. Quantum cosmology seems to suggest that at least our universe is asymmetric in its boundary conditions and thus displays an arrow of time.

Decoherence in the cosmological sense (**b**) refers to the emergence of alternative decoherent sets of histories from the underlying quantum world. Cosmological decoherence is also understood as a general mechanism, whereby the familiar world of our experience—the quasi-classical domain of classical physics with its familiar equations of motion—emerges from some deeper quantum level. Cosmological decoherence aims to explain the emergence of (classical and relativistic) space-times. Thus space-time becomes an emergent phenomenon from a more fundamental quantum level. The macro-world of classical physics is 'quasi-classical' because quantum correlations and quantum superpositions do not disappear but become virtually invisible. They sink below the level of the measurable. The quasi-classical domains consist of 'a branching set of alternative decohering histories' [29, p. 425] but they are 'quasi-classical' because quantum correlations never disappear entirely; they simply become unobservable and no longer affect the motion of the trajectories [30, §6]. According to this approach probabilities can meaningfully be assigned only to

---

[5]Modern cosmology suggests that talk of 'our' universe should be distinguished from talk of 'the' universe, which may harbour parallel universes (multiverse) [10]. According to Linde's inflationary eternal universal model 'our' universe is only one of many universes, which are born and die as islands emerging from a quantum foam. In his latest work, Penrose [53] proposes a cyclic universe, in which never-ending cycles of universes are born one after the other. The lifetime of a universe is an aeon, which stretches from a big bang to a state resembling a heat death, only to start all over again. Penrose's 'cyclic' universe does not respect an overall symmetry but Carroll's model of baby universes does. As Ellis [22] demonstrates, the notion of a multiverse is not without its critics.

[6]Cosmological expansion is problematic as long as there is uncertainty about its uni-directional nature. The second law is seen as problematic because of its statistical nature.

alternative histories, in which the interference effects can be practically ignored.[7] Some possible histories have a negligibly small probability of occurring, whilst others, like the familiar histories of classical systems, have a high probability. Thus, while all histories are equally possible, they are not all equally probable. This was Eddington's point. Eddington also seemed to lean towards the view that there was a master arrow of time.

Decoherence on the cosmological level is responsible for the branching events, the emergence of a set of decoherent histories, as emphasized in some approaches to quantum mechanics. A history is a time-ordered set of properties. Histories are branch-dependent, i.e. they are conditional on past events. Branching events are inherently irreversible, since they select from a set of possible alternative histories one actual history, to which they assign a probability. Branching events replace the act of measurement, since on the cosmological level there are neither observers nor measuring devices.

> We should be able to deal (...) with the classical behavior of the Moon whether or not any observer is looking at it [30, p. 3346].

According to the decoherence approach, applied to the universe as a whole, which is treated as a closed system,

> (s)ets of alternative histories consist of *time sequences* of exhaustive sets of alternatives. A history is a particular sequence of alternatives, abbreviated $[P_\alpha] = (P_{\alpha 1}^1(t_1), P_{\alpha 2}^2(t_2), ... P_{\alpha n}^n(t_n))$ [29, p. 432, italics in original] [40, Appendix].

If the notion of cosmological decoherence plays a significant part in the emergence of 'classical' trajectories in space-time, the notion of irreversiblity will be equally important. It was noted above that environmental decoherence is a form of measurement that occurs in the absence of observers and that irreversibility is the hallmark of measurement situations [29, §11] [30, p. 3376] [31, p. 315] [59, p. 100]. In cosmological decoherence the measurement act is replaced by branching events, by which past records and quasi-classical domains are established. The fact that histories and decoherence are branch-dependent and thus conditional on what happened in their past—their past records—means that these histories are practically irreversible. The emergence of quasi-classical domains of our experience can therefore be understood as an indicator of the arrow of time. Classical cosmology shows that arrows of time arise when initial and final conditions differ, because the initial entropy of the universe is much smaller than the final entropy. Cosmological decoherence provides a mechanism, whereby irreversible classical domains arise from the prior quantum conditions of the universe. It bestows an arrow of time on the classical world; this seems to be the case in our universe.

---

[7]Gell-Mann and Hartle take a rather pragmatic approach to probabilities. 'Probabilities for histories need be assigned by physical theory only to the accuracy to which they are used. (...) We can therefore conveniently consider approximate probabilities, which need obey the rules of the probability calculus only up to some standard of accuracy sufficient for all practical purposes. In quantum mechanics (...) it is likely that only by this means can probabilities be assigned to interesting histories at all' [29, §2, italics in original]. Goldstein [38, p. 14] suggests that their notion of probability is, in effect, a typicality notion.

## 5.5.3 A Master Arrow of Time?

This chapter has proposed that local and cosmic arrows of time are to be conceived, in application of Eddington's and Wheeler's epistemological views, as theoretical constructions (inferences) from available criteria. The passage and the arrows of time cannot be perceived directly, and hence have to be inferred from the empirical phenomena observed in the universe. One astonishing phenomenon is the parallelism of the arrows of time. They are uni-directional. In theory some of these arrows are reversible, in practice they are irreversible. This parallelism of the many arrows of time raises the question whether they have a common origin or whether one arrow is more fundamental than the others. This is the question of the master arrow of time.

From a physical point of view, some arrows would not qualify as master arrows. The psychological arrow is certainly less important, since it lacks regularity and invariance. The historical arrow is too dependent on contingent factors and may not be true of the whole universe.

The afore-mentioned model of the Evolving Block Universe, as developed by cosmologist G. Ellis, shares with the early Eddington the view that the entropic arrow serves as a master arrow of time.

> The other arrows derive from the global master arrow of time resulting from the universe's early expansion from an initial singularity in an Evolving Block Universe. The arrow of time at the start is the time direction pointing away from the initial singularity towards the growing boundary of spacetime; this then remains the direction of time at all later times [23, p. 248].

According to the Evolving Block Universe model, the Big Bang lays the foundation to all the other arrows so that the entropic arrow becomes the master arrow of time.

The existence of the many temporal arrows and their uni-directional nature is undoubted. But the question of the parallelism of the many arrows of time is separate from the question of the master arrow of time. In a similar vein, the fact of the evolution of species was accepted before Darwin explained their origin through the hypothesis of natural selection. Some arrows, such as causal, cosmic and thermodynamic arrows, are far older than biological and psychological arrows of time. On an analogy with evolutionary history, the differential emergence of temporal arrows strongly suggests that there is no master arrow of time. Not all arrows can be reduced to the entropic arrow: the expansion of the universe and the evolution of biological systems seem to run counter to the 'universal' increase in entropy. The evolution of the universe has led to its visible order (constellation of galaxies) even though the overall entropy of the universe is taken to have increased. The causal arrow is not directly linked to the thermodynamic arrow, for causal processes may lead to both increase (causing disorder) and decrease (creating order) in entropy, at least in the local causal field. Some arrows of time keep a tighter link with the thermodynamic arrow of time—for instance the radiative arrow and the cosmic arrows of time. There may be good reasons to think—as some have argued [71]—that the radiative arrow can be reduced to the thermodynamic arrow. The expansion of the universe is either seen as driving the cosmic arrow (Gold universe) or the cosmic arrow drives the expansion of the universe (Penrose, Ellis). Certain cosmological models allow the expansion of the universe to come to a halt and switch over to a contraction to form closed time-like curves. But even in a contracting universe, the entropy would continue to increase; for what changes is the occupied phase space itself but not the individual processes, which take place within it.[8] Rather than speaking of a master arrow of time, it may be more appropriate to adopt the notion of multi-fingered time: that there are mechanisms provided by the universe (expansion, cosmological decoherence, the Second Law), which deliver multiple criteria to infer arrows of time. They emerge at different stages of the evolution of the universe, and are not equally important or reducible to the entropic arrow.

Ultimately, however, both the origin and the existence of arrows of time are still unresolved questions. Tracing the arrows of time has turned out to be a difficult task. This chapter has discussed some theoretical accounts—both their kinematic and dynamic aspects—to explain the existence of temporal arrows. In the author's view it is more plausible to adopt a dynamic view of the universe—with multiple arrows of time—than a static Block Universe.

## 5.6   Conclusion

Are there arrows of time? Some criteria have been considered to suggest that both a global and local arrows could be inferred from available criteria. Both Eddington and Wheeler insisted on the role of inferences from pointer readings (Eddington) or information in terms of bits (Wheeler) to construct a model of the physical universe. One may not want to embrace the operationalism inherent in these claims. For Wheeler, for instance, reality is built from acts of observer-participancy. Eddington claims, on several occasions, that physical quantities are to be defined by measuring operations [18, Chaps. VIII, XIV] [20, Chap. V, §1]. But I wish to propose, with respect to the arrows of time, that the Eddington-Wheeler epistemological approach is the right one. Attempts to identify the arrow of time with particular physical processes have led astray. But there are numerous criteria which allow us to infer the anisotropy of time.

For a long time now two arguments have tempted scientists and philosophers alike to embrace the notion of a static Block Universe: the relativity of simultaneity and the time-reversal invariance of the fundamental laws. The argument ignores the many physical criteria, which point to a more dynamic view of local and global systems. The anisotropy of time is not directly observable. It must be inferred from available criteria. These criteria justify us in regarding arrows of time as conceptual consequences from the available criteria.

---

[8]Consider an analogy: if the universe were a very large billiard table, the borders of which slowly began to retract, the balls would retain their normal trajectories and the players would notice nothing amiss.

# References

1. Aiello, M., Castagnino, M., Lombardi, O.: The arrow of time: from universe time asymmetry to local irreversible processes. Found. Phys. **38**, 257–292 (2008)
2. Albert, D.Z.: Time and Chance. Cambridge (Mass.), Harvard UP, London (2000)
3. Brandenberger, R.: Do we have a theory of early universe cosmology? Stud. Hist. Philos. Mod. Phys. **46**, 109–121 (2013)
4. Boltzmann, L.: The Second Law of Thermodynamics. In: Boltzmann, L. (ed.) (1974), pp. 13–32 (1886)
5. Boltzmann, L.: Reply to Zermelo's Remarks on the Theory of Heat. 1966, pp. 218–228 (1896)
6. Boltzmann, L.: On Zermelo's Paper 'On the Mechanical Explanation of Irreversible Processes'. L. Boltzmann 1966, 238–245 (1897)
7. Boltzmann, L.: Vorlesungen über Gas-Theorie, II. Johann Ambrosius Barth, Leipzig (1898/1923)
8. Boltzmann, L.: Kinetic theory, vol. 2. In: Brush (ed.), Irreversible Processes. Pergamon Press, Oxford, London (1966)
9. Boltzmann, L.: Theoretical physics and philosophical problems. In: McGuinness, B. (ed.) Selected Writings. D. Reidel Publishing Company, Dordrecht, Holland (1974)
10. Carroll, S.: From Eternity to Here. OneWorld, Oxford (2010)
11. Castiagnino, M., Lara, L., Lombardi, O.: The direction of time: from the global arrow to the local arrow. Int. J. Theor. Phys. **42/10**, 2487–2504 (2003)
12. Earman, J.: An attempt to add a little direction to 'the problem of the direction of time'. Philos. Sci. **41**(1), 15–47 (1974)
13. Earman, J.: World Enough and Space-Time. Cambridge (Mass.), MIT Press, London (1989)
14. Earman, J.: What time reversal invariance is and why it matters. Int. Stud. Philos. Sci. **16**(3), 245–264 (2002)
15. Earman, J.: The 'past hypothesis': not even false. Stud. Hist. Philos. Mod. Phys. **37**, 399–430 (2006)
16. Earman, J.: How determinism can fail in classical physics and how quantum mechanics can (sometimes) provide a cure. Philos. Sci. **75**(5), 817–829 (2008)
17. Eddington, A.S.: The Theory of Relativity and its Influence on Scientific Thought. Clarendon Press, Oxford (1922)
18. Eddington, A.S.: The Nature of the Physical World. Cambridge University, Cambridge (1932/1928)
19. Eddington, A.S.: New Pathways in Science. Cambridge University Press, Cambridge (1935)
20. Eddington, A.S.: The Philosophy of Physical Science. Cambridge University Press, Cambridge (1939)
21. Feyman, R.: The Character of Physical Law. Penguin Books Press (1965)
22. Ellis, F.R.: Does the Multiverse Really Exist? Scientific American, pp. 18–23 (August 2011)
23. Ellis, G.F.R.: The arrow of time and the nature of spacetime. Stud. Hist. Philos. Mod. Phys. **44**, 242–262 (2013)
24. Frigg, R.: A field guide to recent work on the foundations of statistical mechanics. In: Rickles, D. (ed.) The Ashgate Companion to Contemporary Philosophy of Physics, pp. 99–196. Ashgate, London (2008)
25. Frigg, R.: Typicality and the approach to equilibrium in boltzmannian statistical mechanics. Philos. Sci. (Supplement) **76**, 997–1008 (2009)
26. Frigg, R.: Why typicality does not explain the approach to equilibrium. In: Surez, M. (ed.) Probabilities, Causes and Propensities in Physics, pp. 77–93. Springer, Synthese Library, Heidelberg (2011)
27. Frigg, R., Werndl, Ch.: Explaining thermodynamic-like behavior in terms of epsilon-ergodicity. Philos. Sci. **78**, 628–652 (2011)
28. Frigg, R., Werndl, Ch.: Demystifying typicality. Philos. Sci. **79**, 917–929 (2012)

29. Gell-Mann, M., Hartle, J.B.: Quantum mechanics in the light of quantum cosmology. In: Zurek, W. (ed.) Complexity, Entropy and the Physics of Information, pp. 425–465. Addison-Wesley (1990)
30. Gell-Mann, M., Hartle, J.B.: Classical equations for quantum systems. Phys. Rev. D **47**(8), 3345–3382 (1993)
31. Gell-Mann, M., Hartle, J.B.: Time symmetry and asymmetry in quantum mechanics and quantum cosmology. In: Halliwell, J.J., Prez-Mercader, J., Zurek, W.H. (eds.) Physical Origins of Time Asymmetry, pp. 311–345. Cambridge University Press, Cambridge (1994)
32. Giulini, D., Joos, E., Kiefer, C., Kupsch, J., Stamatescu, I.-O., Zeh, H.D.: Decoherence and the Appearance of a Classical World in Quantum Theory. Springer, Berlin/Heidelberg/New York (1996)
33. Gold, T.: The arrow of time. Am. J. Phys. **30**, 403–410 (1962)
34. Gold, T.: Cosmic processes and the nature of time. In: Colodny, R.G. (ed.) Mind and Cosmos, pp. 311–329. University of Pittsburgh Press (1966)
35. Gold, T.: The world map and the apparent flow of time. In: Gal-Or, B. (ed.) Modern Developments in Thermodynamics, pp. 63–72. Wiley, New York (1974)
36. Goldstein, Sh, Lebowitz, J.L.: On the (Boltzmann) entropy of non-equilibrium systems. Physica D **193**, 53–66 (2004)
37. Goldstein, S.: Projection Postulate. In: Greenberger, D., Hentschel, K., Weinert, F. (eds.) Compendium of Quantum Physics, pp. 499–501. Springer, Berlin/Heidelberg (2009)
38. Goldstein, S.: Typicality and notions of probability in physics. In: Ben-Menahem, Y., Hemmo, M. (eds.) Probability in Physics, pp. 59–71. Springer, Heidelberg (2012)
39. Halliwell, J.J., Prez-Mercader, J., Zurek, W.H. (eds.): Physical Origin of Time Asymmetry. Cambridge University Press, Cambridge (1994)
40. Hartle, J.: The quasiclassical realms of this quantum universe. Found. Phys. **41**, 982–1006 (2011)
41. Hawking, St. W.: The no boundary condition and the arrow of time. In: Halliwell, J.J., Prez-Mercader, J., Zurek, W.H. (eds.) Physical Origin of Time Asymmetry, pp. 346–357. Cambridge University Press, Cambridge (1994)
42. Kittel, Ch., Kroemer, H.: Thermal Physics. W. H. Freeman and Company, New York (1980)
43. Kupervasser, O., Nikolić, H., Zlatić, V.: The universal arrow of time. Found. Phys. **42**, 1165–1185 (2012)
44. Landsberg, P.T.: Introduction. In: Landsberg, P.T. (ed.) The Enigma of Time, pp. 1–30. Bristol, Adam Hilger Ltd (1982)
45. Landsberg, P.T.: A matter of time. In: Landsberg, P.T. (ed.) The Enigma of Time, pp. 60–85. Bristol, Adam Hilger Ltd (1982)
46. Landsberg, P.T. (ed.): The Enigma of Time. Bristol, Adam Hilger Ltd (1982)
47. Lebowitz, J.L.: Time's Arrow and boltzmann's entropy. In: Halliwell, J.J., et al. (eds.) Physical Origin of Time Asymmetry, pp. 131–146. Cambridge University Press, Cambridge (1994)
48. Leff, H.S.: Entropy, Its language and interpretation. Found. Phys. **37**, 1744–1766 (2007)
49. Lehmkuhl, D.: On time in space-time. Philosophia Naturalis **49**(2), 225–237 (2012)
50. Penrose, R.: Big bangs, black holes and 'time's arrow'. In: Flood, R., Lockwood, M. (eds.) The Nature of Time, pp. 36–62. Basil Blackwell, Oxford (1986)
51. Penrose, R.: The Emperor's New Mind. Vintage, Oxford/New York (1989)
52. Penrose, R.: The Road to Reality. Vintage Books, London (2005)
53. Penrose, R.: Cycles of Time. The Bodley Head, London (2010)
54. Price, H.: Time's Arrow and Archimedes' Point. Oxford University Press, Oxford (1996)
55. Price, H.: Boltzmann's time bomb. Br. J. Philos. Sci. **53**, 83–119 (2002)
56. Rugh, S.E., Zinkernagel, H.: On the physical basis of cosmic time. Stud. Hist. Philos. Mod. Phys. **40**, 1–19 (2009)
57. Schlegel, R.: Time and the Physical World. Dover Publications (1968)
58. Schulman, L.: Time's Arrow and Quantum Measurement. Cambridge University Press, Cambridge (1997)

59. Schlosshauer, M.: Decoherence and the Quantum-to-Classical Transition, pp. 511–523, vol. 99, Part VIII. Springer Proceedings in Physics, Heidelberg/London/New York (2008)
60. Sklar, L.: Physics and Chance. Cambridge University Press, Cambridge (1993)
61. Stöckler, H. (ed.): Taschenbuch der Physik. Verlag Harri Deutsch, Frankfurt a. M (2000)
62. Uffink, J.: Bluff your way in the second law of thermodynamics. Stud. Hist. Philos. Mod. Phys. **32**, 305–394 (2001)
63. Volchan, S.B.: Probability as typicality. Stud. Hist. Philos. Mod. Phys. **38**, 801–814 (2007)
64. Vranas, P.B.M.: Epsilon-ergodicity and the success of equilibrium statistical mechanics. Philos. Sci. **65**(4), 688–708 (1998)
65. Weinert, F.: The Scientist as Philosopher. Springer, Heidelberg/Berlin/New York (2004)
66. Weinert, F.: Copernicus, Darwin and Freud. Wiley (2009)
67. Weinert, F.: The March of Time. Springer, Heidelberg/Berlin/New York (2013)
68. Werndl, Ch.: Justifying typicality measures of Boltzmannian statistical mechanics and dynamical systems. Stud. Hist. Philos. Mod. Phys. **44**, 470–479 (2013)
69. Wheeler, J.A.: Three-dimensional geometry as a carrier of information about time. In: Gold, T. (ed.) The Nature of Time, pp. 90–110. Ithaca (New York), Cornell UP (1967)
70. Wheeler, J.A.: It from bit. In: Wheeler, J.A. (ed.) At Home in the Universe, pp. 295–311. Springer, Heidelberg/Berlin/New York (1996)
71. Zeh, H.D.: The Physical Basis of the Direction of Time. Springer, Heidelberg/Berlin/New York (1992)
72. Zurek, W.H.: Preferred sets of states, predictability, classicality, and environment-induced decoherence. In: Halliwell, J.J., Prez-Mercader, J., Zurek, W.H. (eds.) Physical Origin of Time Asymmetry, pp. 175–212. Cambridge University Press, Cambridge (1994)

# Chapter 6
# Constructor Theory of Information

Chiara Marletto

The word 'information' is connected with a rich undergrowth of resilient, interrelated problems, whose origin resides in the apparently contradictory nature of the properties that information displays. On the one hand, information seems to be central to fundamental physics, for many reasons—already envisaged by Wheeler in his 'it from bit' programmatic article [14]. To mention a few, information is at the foundation of the theories of preparation and measurement; besides there are (isolated) laws which refer directly to information and link it to exact, fundamental quantities—such as Beckenstein's formula for the entropy of a black hole [1]. Moreover, the quantum theory of computation has showed that the properties of information-processing devices, such as computers, strongly depend on the underlying dynamical laws. On the other hand, information does not have the status of other physical quantities that physics resorts to in order to describe reality: it is not expressible as a function of the fields on spacetime only, nor as a quantum-mechanical observable. Moreover, it behaves rather like an abstraction: laws about information do not refer explicitly to physical systems (this is the *substrate independence of information*); also, information can be copied from one physical system to any other and still preserves all its properties, irrespective of their physical details (this is what we call *interoperability*).

Indeed, from time to time it has been claimed that information is only a manner of discourse—something that at most can have a pedagogical role in physics. Given these contradictory features, what can it possibly mean to formulate a theory of information *within fundamental physics*?

A new perspective on this issue has emerged of late, reformulating the problem via an elegant and deep side-stepping. According to this new approach—which, slightly to jump ahead, goes under the name "constructor theory of information", [4, 6, 7]—the problem is *not* to formulate a physical theory providing a defini-

C. Marletto (✉)
Materials Department, University of Oxford, Oxford, UK
e-mail: chiara.marletto@gmail.com

© Springer International Publishing Switzerland 2017
I.T. Durham and D. Rickles (eds.), *Information and Interaction*,
The Frontiers Collection, DOI 10.1007/978-3-319-43760-6_6

tion of the entity called "information" in terms of underlying known physical concepts; instead, it is to formulate a physical theory of the *regularities in the laws of physics* required for there to exist what has been vaguely referred to as "information". Likewise, the problem thermodynamics addressed is *not* that of defining a quantity called "caloric", as it was initially thought; but to express the properties that such a concept—or, rather, its improved versions, i.e., work and heat—implicitly require. As we shall see, the interesting point of the constructor theory of information is that it is capable of expressing the properties of the physical laws permitting information *exactly*, without resorting to any approximations.

As an example of the regularities associated with the concept of "information", consider interoperability—as we said above, this is the property of information being *copiable* from one physical instantiation (e.g. transistors in a computer) to a different one (e.g. the flags of an air-traffic controller). This is a regularity displayed by the laws of physics of our universe, which is taken for granted by the current theories of information and computation. However, one could imagine laws that did not have it—under which 'information' (as we informally conceive it) would not exist. For example, consider a universe where there existed two sectors A and B, each one allowing copying-like interactions between media inside it, but such that no copying interactions were allowed between A and B. There would be no 'information' (as informally conceived) in the composite system of the two sectors. This is an example of how whether or not information can exist depends on the presence of certain regularities in the laws of physics.

Another example is found in the properties assumed by Shannon's theory of information, [11]. It *requires* physical systems to have *distinguishable states*: these are needed for communication to be possible. In fact, in Shannon's (idealised) setting this assumption does not have to be explicitly expressed, because the underlying physics is, in effect, a discrete version of classical mechanics—where any two different states can (in principle) be distinguished with arbitrarily high accuracy. Not only is this false in quantum theory (because of the no-cloning principle, [15]), but, from our present perspective, the problem arises of what 'distinguishability' requires of the physical system whose states are to be distinguished: what kind of interactions must it be capable of undergoing? This question must be answered by considering how the distinguishing process occurs physically. It requires a non-perturbing measurement to happen at some point, which distinguishes two possible states $x$ and $y$ of some physical system (the source), having the following effect in those two cases:

$$
\begin{array}{cccc}
\text{source} & \text{target} & \text{source} & \text{target} \\
x & x_0 & \rightarrow \quad x & x' \\
y & x_0 & \rightarrow \quad y & y'
\end{array}
\qquad (6.1)
$$

where $x_0$ is a blank state of some receptive, target, medium whose interaction with the source is such that at the end of the measurement it instantiates the outcome $x'$ or $y'$. But this does not in fact distinguish message $x$ from message $y$ unless the target states $x'$ and $y'$ are themselves distinguishable. Therefore (6.1), considered as a definition of distinguishability, is recursive and needs a base for the recursion.

The base must still be an operational criterion, since we are seeking an answer to the above question that is independent of any particular theory. (For example, "$x'$ and $y'$ are orthogonal" would not do, as it is special to quantum theory's formalism; we need to specify what it is about orthogonal states that would make them usable for communication in Shannon's sense.) The *constructor theory of information* provides the base, among other things. In so doing, it expresses in an exact, scale-independent way what constraints the laws of physics must obey in order for them to instantiate what we have learnt informally to call 'information'. Thus it accommodates the notion of (classical) information within fundamental physics and unifies it exactly with what currently goes under the name of 'quantum information'—the kind of information that is deemed to be instantiated in quantum systems. Rather crucially, the constructor theory of information differs from previous approaches to incorporating information into fundamental physics, e.g. Wheeler's 'it from bit', [14], in that it does not consider information itself as an a priori mathematical or logical concept. Instead, it requires that the nature and properties of information follow entirely from the laws of physics.

Crucial to the formulation of such a theory is a new mode of explanation in fundamental physics: *constructor theory*, [4]. Constructor theory is a new theory of physics intended to improve upon current fundamental theories and explain more of physical reality. Its central idea is to formulate *all laws of physics as statements about what transformations are possible, what are impossible, and why.* This is a sharp departure from the prevailing conception of fundamental physics, which can only predict what *will happen*, given the initial conditions and the laws of motion. For instance, the prevailing conception of fundamental physics aims at predicting the trajectory of a rocket, given its initial conditions and the laws of motion of the universe; instead, in constructor theory the fundamental statements are about where the rocket *could be made to go*, with given resources, under the dynamical laws.

This constructor-theoretic, task-based formulation of science makes new conceptual tools available in fundamental physics, which resort to *counterfactual statements* (i.e., about possible and impossible tasks). Such counterfactual statements are central to expressing the properties of the laws of physics permitting information: that a medium can instantiate information, that two states are distinguishable from one another, or that information can be copied from one system to another, all refer to certain tasks being *possible* under the laws of physics; they cannot possibly be expressed via statements in terms of predictions in the prevailing conception. As Weaver put it, [13], information has a counterfactual nature:

> this word 'information' in communication theory relates not so much to what you do say, as to what you could say.

Let me now recall the foundations of constructor theory. Constructor theory consists of "laws about laws": its laws are *principles*—conjectured laws of nature, in the same sense that the principle of conservation of energy is. It underlies our currently most fundamental theories of physics, such as general relativity and quantum theory, which in this context we call *subsidiary theories*. These are not to be derived from constructor theory; in contrast, constructor theory's principles constrain subsidiary theories, ruling out some of them.

The fundamental principle of constructor theory is that *every physical theory is expressible via statements about what physical transformations, or tasks, are possible, what are impossible, and why.* Therefore, constructor theory requires explanations of the physical world to be formulated in terms of statements about the possibility of transformations, or, more precisely, *tasks*. Physical transformations are defined as physical processes involving two kinds of physical systems, with different roles. One is the *constructor*, whose defining characteristic is that it causes the transformation to occur and remains *unchanged in its ability to cause it again*. The other consists of the subsystems—called the *substrates*—which are transformed from having some physical attribute to having another. Schematically:

$$\text{Input attributes of substrates} \xrightarrow{\text{Constructor}} \text{Output attributes of substrates,}$$

where the constructor and the substrates jointly constitute an isolated system. For example, the catalyst in a chemical reaction approximates a constructor and the chemicals being transformed are its substrates. By *attribute* here one means a set of states of a system in which the system has a certain property according to the subsidiary theory describing it—such as being red or blue. The basic objects of constructor theory are *tasks*. They are the specifications of only the input-output pairs of a transformation, with the constructor abstracted away:

$$\text{Input attributes of substrates} \rightarrow \text{Output attributes of substrates.}$$

Therefore a task A on a substrate **S** is a set:

$$A = \{x_1 \rightarrow y_1, \quad x_2 \rightarrow y_2, ...\},$$

where the $x_1, x_2 \ldots$ and the $y_1, y_2 \ldots$ are attributes of **S**. The set $\{x_i\} = \text{In}(A)$ are the legitimate input attributes of A and the set $\{y_i\} = \text{Out}(A)$ its legitimate output attributes. Tasks can be composed into networks, by serial and parallel composition, to form other tasks.

It is remarkable that this is an explicitly local framework, requiring that individual physical systems have states (and attributes). Indeed, Einstein's principle of locality [9] is an essential principle of constructor theory: *There exists a mode of description such that the state of the combined system* $S_1 \oplus S_2$ *of any two substrates* $S_1$ *and* $S_2$ *is the pair* $(x, y)$ *of the states x of* $S_1$ *and y of* $S_2$, *and any construction undergone by* $S_1$ *and not* $S_2$ *can change only x and not y.* Quantum theory satisfies this principle, as it is explicit in the Heisenberg picture (see [5]).

A constructor is *capable of performing a task* A if, whenever presented with substrates having an attribute in the set of legitimate input attributes $\text{In}(A)$, it delivers them with one of the corresponding attributes in the set of permitted output attributes $\text{Out}(A)$.

A task A is *impossible* if there is a law of physics that forbids it. Otherwise it is *possible*—i.e., there is no limit, short of perfection, on how accurately A could be performed, nor on how well things that are capable of approximately performing it could retain their ability to do so again. However, *no perfect constructors exist in nature*, given our laws of physics. Approximations to them, such as catalysts, enzymes or computers, do make errors and also deteriorate with use. Thus that a task is possible means the laws of nature permit the existence of an approximation to a constructor for that task to any given finite accuracy. The notion of a constructor must therefore be intended as representing the infinite sequence of these approximations.

Note that this is a *deterministic* framework: probabilistic theories can only be approximate descriptions of reality in the world-view of constructor theory. For how probabilities emerge in constructor theory see [10]: the logic mirrors that of how probabilities emerge in unitary quantum theory [3, 12]. For a discussion of the testability of such a theory, see [2].

For present purposes we are interested in what is possible or impossible irrespective of the kind of resources required. So, whenever it is possible to perform the task A in parallel with some task T on some generic, preparable substrate—see [6]—one says that A is *possible with side-effects*, which we shall denote by $A^{\swarrow}$. (The task T is the side-effect).

Subsidiary theories complying to constructor theory must therefore express everything important about the world via statements about the possibility and impossibility of tasks only, *without mentioning constructors*. It is precisely abstracting away the constructor that makes it possible, in the constructor theory of information, to express the regularities associated with the informally conceived notion of (classical) information.

The logic of how the theory is constructed is elegant and simple. The first key step is to express information in terms of computations, not vice-versa as is usually done; and to express computation in terms of tasks—which are primitives and fundamental in constructor theory.

A *reversible computation* $C_\Pi (S)$ is the task of performing, with or without side-effects, a permutation $\Pi$ over some set $S$ of at least two possible attributes of some substrate:

$$C_\Pi (S) = \bigcup_{x \in S} \{x \to \Pi(x)\}.$$

By a 'reversible computation' $C_\Pi$ is meant a logically reversible, i.e., one-to-one, task; but the process that implements it may be physically irreversible, because of the possibility of side-effects.

A *computation variable* is a set $S$ of two or more possible attributes for which $C_\Pi^{\swarrow}$ for all permutations $\Pi$ over $S$, and a *computation medium* is a substrate with at least one computation variable. For instance, a quantum bit in any two non-orthogonal states is an example of a computation medium: these two states can be swapped by a unitary transformation, despite their not being distinguishable by a single-shot measurement [7]. Thus, the elements of a computation variable need not be distinguishable.

Next, one introduces the notion of an *information medium*. This requires computations involving two instances of the same substrate **S** to be considered. The *cloning
task* for a set $S$ of possible attributes of **S** is

$$R_S(x_0) = \bigcup_{x \in S} \{(x, x_0) \to (x, x)\} \tag{6.2}$$

on $\mathbf{S} \oplus \mathbf{S}$, where $x_0$ is some fixed (independent of $x$) attribute with which it is possible
to prepare **S**. A set $S$ is *clonable* if $R_S$ is possible (with or without side-effects) for
some such $x_0$.

An *information variable* is a clonable computation variable. An *information
attribute* is one that is a member of an information variable, and an *information
medium* is a substrate that has at least one information variable.

We are now in the position to provide a base for the definition of 'distinguishable'.
A variable $X$ of a substrate **S** is distinguishable if the task

$$\bigcup_{x \in S} \{(x, x_0) \to (\psi_x)\} \tag{6.3}$$

is possible with or without side-effects, where the $\{\psi_x\}$ constitute an information
variable. Since "information variable" is entirely defined in terms of the tasks of
permuting and cloning $X$ being possible, and *not* in terms of information, nor in terms
of distinguishability, this is a base for the operational criterion at the foundations of
Shannon's theory of information (see Eq. 6.1), thus rooting the latter into physics.
For instance, with the tools we have just introduced one can say that any theory
of physics with information variables can support communication in the sense of
Shannon's.

Another thing that is usually fuzzily defined, but can be expressed exactly in this
new framework, is the notion of a system containing information. Specifically, a
substrate **S** *instantiates information* if it is in a state belonging to some attribute in
some information variable $X$ of **S** and if it could have been given any of the other
attributes in $S$. The constructor-theoretic mode of explanation has allowed these
properties to be expressed as *exact, intrinsic* properties of the substrate by switching
to statements about possible/impossible tasks. The prevailing conception, instead,
would require one to express it as a property of the composite system of substrate
and constructor.

One can then conjecture *principles about information media*, that have elegant
expressions in terms of possible and impossible tasks: these are the regularities
tacitly assumed by theories of computation and information. A cardinal principle is
the *interoperability principle*: *The combination of two substrates with information
variables $X_1$ and $X_2$ is a substrate with information variable $X_1 \times X_2$*, where $\times$
denotes the Cartesian product of sets. The meaning of this principle is that information
can be copied from any kind of information medium to any other kind. This property
cannot possibly be stated in the prevailing conception of fundamental physics, but
has an elegant, terse expression in constructor theory. Similar expressions are found

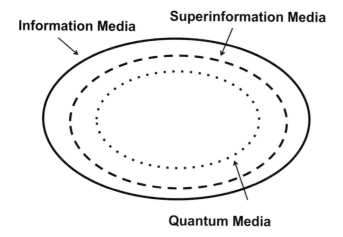

**Fig. 6.1** Classification of information media in the constructor theory of information

for the other principles of information, see [7], which represent the physical laws that underlie the existence of (classical) information.

This construction provides a purely constructor-theoretic notion of what was previously called classical information. However, it has emancipated it from classical physics: it is expressed in a totally general framework.

Since quantum theory is so much deeper and general than classical, one might then think that it is necessary to conjecture some additional law, declaring some counter-intuitive tasks to be possible, in order to incorporate in this framework what we currently call quantum information—i.e., the kind of information that can be instantiated in systems obeying quantum theory. Quite surprisingly, it turns out that one can define a new kind of medium, called *super-information medium*, obeying a simple constraint about a task being *impossible*. Specifically, a superinformation medium is an information medium that has two information variables, the union of which is *not* an information variable. From this one constraint one can derive all the most distinctive, qualitative properties of quantum information [7]:

- Existence of pairs of attributes that cannot be cloned;
- Irreducible perturbation caused by measurement;
- Unpredictability of the outcome of certain experiments;
- Locally inaccessible information in composite systems, as in entangled systems (on the additional requirement of reversibility).

These properties are naturally expressible in constructor-theoretic terms because they are all about certain tasks being possible or impossible. This provides a unifying framework where quantum and classical information are related to one another, *exactly*, in a way that was not previously hinted at: it turns out that quantum information is an instance of superinformation, see Fig. 6.1.

Thus, the constructor theory of information explains the regularities in physical systems that can instantiate information and quantum information, characterising them in an exact, scale-independent way. It also introduces a generalisation of quantum systems instantiating quantum information: superinformation media. These may have the same information-processing capabilities as quantum information media, but need not obey quantum theory. Indeed, the constructor theory of information also provides a framework, independent of quantum theory, where one can investigate information under a broad range of theories (including 'postquantum' theories) if they obey the principles. It can be conjectured that the putative successor of quantum theory is one that allow some kind of super information media, in the set of those allowing for probabilities (as in [10]). In this respect, the construction described here challenges the worldview that quantum theory is 'random' and 'non-local': constructor theory demands subsidiary theories to be local and deterministic.

This cornucopia of constructor-theoretic consequences has originated from a single move: switching to the mode of explanation via statements about possible/impossible tasks—i.e., taking counterfactual (rather than factual) statements as fundamental. This fact should not come as a surprise. After all, as Eddington once said [8]:

> We need scarcely add that the contemplation in natural science of a wider domain than the actual leads to a far better understanding of the actual.

**Acknowledgements** The author thanks David Deutsch for illuminating discussions as well as for helpful comments on earlier drafts. This publication was made possible through the support of a grant from the Templeton World Charity Foundation. The opinions expressed in this publication are those of the author and do not necessarily reflect the views of Templeton World Charity Foundation.

# References

1. Beckenstein, J.: Black-hole thermodynamics. Phys. Today **33**(1), 24–31 (1980)
2. David, D.: The logic of experimental tests, particularly of everettian quantum theory. In: Studies in History and Philosophy of Science Part B: Studies in History and Philosophy of Modern Physics (2016)
3. Deutsch, D.: Quantum theory of probability and decisions. Proc. R. Soc. london **A455**, 3129–3137 (1999)
4. Deutsch, D.: Constructor theory. Synthese **190**(18), 4331–4359 (2013)
5. Deutsch, D., Hayden, P.: Information flow in entangled quantum systems. Proc. R. Soc. London **A456**, 1759–1774 (2000)
6. Deutsch, D., Marletto, C: Reconstructing reality. New Sci. (24 May 2014)
7. Deutsch, D., Marletto, C.: Constructor theory of information. Proc. R. Soc. A **471**, 20140540 (2015)
8. Eddington, A.: The nature of the physical world (1929)
9. Einstein, A.: Quoted in Albert Einstein: philosopher, scientist. In: Schilpp, P.A. (ed.) Library of Living Philosophers, Evanston, 3rd ed., p. 85, (1970) (1949)
10. Marletto, C.: Constructor theory of probability. Proc. R. Soc. A **472**, 2192 (2016) doi:10.1098/rspa.2015.0883
11. Shannon, C.E.: A mathematical theory of communication. Bell Syst. Tech. J. **27**(379–423), 623–656 (1948)

12. Wallace, D.: Everettian rationality: defending Deutsch's approach to probability in the Everett interpretation. Stud. History Philos. Modern Phys. **34**, 415–439 (2003)
13. Weaver, W., Shannon, C.E., Weaver, W.: The Mathematical Theory of Communication. Urbana: University of Illinois Press (1949)
14. Wheeler, J.A.: Information, physics, quantum: the search for links. In: Proceedings III International Symposium on Foundations of Quantum Mechanics, pp. 354–368. Tokyo (1989)
15. Wootters, W., Zurek, W.: A single quantum cannot be cloned. Nature **299**, 802–803 (1982)

# Chapter 7
# On Participatory Realism

Christopher A. Fuchs

**Abstract** In the *Philosophical Investigations*, Ludwig Wittgenstein wrote, "'I' is not the name of a person, nor 'here' of a place, …. But they are connected with names. … [And] it is characteristic of physics not to use these words." This statement expresses the dominant way of thinking in physics: Physics is about the impersonal laws of nature; the "I" never makes an appearance in it. Since the advent of quantum theory, however, there has always been a nagging pressure to insert a first-person perspective into the heart of physics. In incarnations of lesser or greater strength, one may consider the "Copenhagen" views of Bohr, Heisenberg, and Pauli, the observer-participator view of John Wheeler, the informational interpretation of Anton Zeilinger and Časlav Brukner, the relational interpretation of Carlo Rovelli, and, most radically, the QBism of N. David Mermin, Rüdiger Schack, and the present author, as acceding to the pressure. These views have lately been termed "participatory realism" to emphasize that rather than relinquishing the idea of reality (as they are often accused of), they are saying that reality is *more* than any third-person perspective can capture. Thus, far from instances of instrumentalism or antirealism, these views of quantum theory should be regarded as attempts to make a deep statement about the nature of reality. This paper explicates the idea for the case of QBism. As well, it highlights the influence of John Wheeler's "law without law" on QBism's formulation.

John Archibald Wheeler's writings were a tremendous influence on the point of view on quantum mechanics we now call QBism [24], developed by the author [16, 17, 22, 26] and colleagues Caves [11–14], Peres [18, 19], Appleby [1–7], Stacey [50–52], von Baeyer [54], Mermin [32, 33, 37–47], and most especially Rüdiger

C.A. Fuchs (✉)
Department of Physics, University of Massachusetts Boston,
100 Morrissey Boulevard, Boston, MA 02125, USA
e-mail: qbism.fuchs@gmail.com

C.A. Fuchs
Max Planck Institute for Quantum Optics, Hans-Kopfermann-Strasse 1,
85748 Garching, Germany

© Springer International Publishing Switzerland 2017
I.T. Durham and D. Rickles (eds.), *Information and Interaction*,
The Frontiers Collection, DOI 10.1007/978-3-319-43760-6_7

Schack [20, 21, 23, 27–29, 31, 34, 49].[1] It is true that the term was initially an abbreviation for Quantum Bayesianism, but long before Bayesianism was brought into the mix of my own thinking (I won't speak for the others), there were John Wheeler's exhortations on "law without law" filling my every sleepless night. A typical something Wheeler would write on the subject would sound like this,

> "Law without law": It is difficult to see what else than that can be the "plan" for physics. It is preposterous to think of the laws of physics as installed by a Swiss watchmaker to endure from everlasting to everlasting when we know that the universe began with a big bang. The laws must have come into being. Therefore they could not have been always a hundred percent accurate. That means that they are derivative, not primary. Also derivative, also not primary is the statistical law of distribution of the molecules of a dilute gas between two intersecting portions of a total volume. This law is always violated and yet always upheld. The individual molecules laugh at it; yet as they laugh they find themselves obeying it. … Are the laws of physics of a similar statistical character? And if so, statistics of what? Of billions and billions of acts of observer-participancy which individually defy all law? … [Might] the entirety of existence, rather than [be] built on particles or fields or multidimensional geometry, [be] built on billions upon billions of elementary quantum phenomena, those elementary acts of observer-participancy?

Roger Penrose called the idea "barely credible,"[48] but there was something about it that attracted me like nothing else in physics: Right or wrong—or more likely, not well-defined enough to be *right* or *wrong*!—Wheeler's thinking was the very reason I pursued the study of physics in the first place. In college and graduate school, I would read every piece of Wheeler's writing on the subject that I could find, no matter how repetitious the task became.[2]

One thing that is absolutely clear in Wheeler's writings is that the last thing he was pursuing was an *instrumentalist* understanding of quantum theory. He thought that quantum theory was the deepest statement on *nature* and *reality* ever discovered by physics. Thus, he would play little dialogues with himself like this one: "The strange

---

[1] A further important review of QBism by an outsider can be found in [53].

[2] The exact corpus of my readings was this: [55–110, 112–114],

> [111] I want you and Einstein to jolt the world of physics into an understanding of the quantum because the quantum surely contains—when unraveled—the most wonderful insight we could ever hope to have on how this world operates, something equivalent in scope and power to the greatest discovery that science has ever yet yielded up: Darwin's Evolution.
> You know how Einstein wrote to his friend in 1908, "This quantum business is so incredibly important and difficult that everyone should busy himself with it." … Expecting something great when two great minds meet who have different outlooks, all of us in this Princeton community expected something great to come out from Bohr and Einstein arguing the great question day after day—the central purpose of Bohr's four-month, spring 1939 visit to Princeton—I, now, looking back on those days, have a terrible conscience because the day-after-day arguing of Bohr was not with Einstein about the quantum but with me about the fission of uranium. How recover, I ask myself over and over, the pent up promise of those long-past days? Today, the physics community is bigger and knows more than it did in 1939, but it lacks the same feeling of **desperate** puzzlement. I want to recapture that feeling for us all, even if it is my last act on Earth.

necessity of the quantum as we see it everywhere in the scheme of physics comes from the requirement that—via observer-participancy—the universe should have a way to come into being? If true—and it is attractive—it should provide someday a means to <u>derive</u> quantum mechanics from the requirement that the universe must have a way to come into being." Indeed, he was fond of quoting Niels Bohr, who in his last interview (the day before he died), said, "I felt … that philosophers were very odd people who really were lost, because they have not the instinct that it is important to learn something and that we must be prepared really to learn something of very great importance …." Instrumentalists don't learn lessons about nature! They just try to predict what is coming next as best they can.

So, it is with some dismay and consternation that I hear over and over that QBism is an instrumentalist account of quantum theory. For instance, in Jeffrey Bub's recent book *Bananaworld: Quantum Mechanics for Primates* [9],[3] he writes,

> Another approach to the conceptual problems of quantum mechanics is the "quantum Bayesianism" or "QBism" of Christopher Fuchs and Rüdiger Schack. David Mermin is a recent convert. … On this view, all probabilities, are understood in the subjective sense as the personal judgements of an agent, based on how the external world responds to an action by the agent. Then there's no problem in explaining how quantum probabilities of "what you'll obtain if you measure" are related to classical ignorance of "what's there," because all probabilities are personal bets about what you'll find if you look. This approach seems to be straightforwardly instrumentalist from the start: a theory is a useful tool, a predictive device, based on a summary of how the world responds to an agent's actions.

When I read things like this, I find myself wanting to say, "[P]hilosophers [are] very odd people who really [are] lost, because they have not the instinct that it is important to learn something and that we must be prepared really to learn something of very great importance …" What is at stake with quantum theory is the very nature of reality. Should reality be understood as something completely impervious to our interventions, or should it be viewed as something responsive to the very existence of human beings? Was the big bang all and the whole of creation? Or is creation going on all around us all the time, and we ourselves are taking part in it? No philosophical predisposition to instrumentalism led inexorably to QBism: Quantum theory itself threw these considerations before us!

The bulk of this paper is made of correspondence, in the vein of a couple of my previous contributions to the literature [25, 30], where I try to set the record straight on this. This correspondence was prompted predominantly by some early drafts of Adán Cabello's paper, "Interpretations of Quantum Theory: A Map of Madness," [10] the final form of which is posted here: http://arXiv.org/abs/1509.04711. Initially Adán labelled the views on quantum mechanics of the Copenhagen School, çaslav Brukner and Anton Zeilinger, Asher Peres, Carlo Rovelli, John Wheeler, and the QBism of David Mermin, Rüdiger Schack, and myself, as all "non-realist." [See Adán's paper for appropriate citations to these various views.] This incensed me so, bringing back

---

[3]But other similar accounts abound. Here is another choice one, from Ladyman and Ross's book *Everything Must Go: Metaphysics Naturalized* [36]: "According to [Fuchs and Peres] the quantum state of a system is just a probability distribution in disguise. This is an instrumentalist approach that is metaphysically unhelpful."

waves of emotion over having QBism called instrumentalism, solipsism, anti-realism, mysticism, and even psychology, that I urged the term "participatory realism" to try to capture what is in common among these views, while calling attention outright that they are all *realist* takes on the task of physics. It also seemed a worthy tribute to John Wheeler, as I thought he captured the appropriate sentiment with his phrase "participatory universe"—it's as full-blown a notion of reality as anyone could want, recognizing only that the users of quantum mechanics have their part of it too. That's not less reality, that's more. Ultimately Adán adopted the term "Type-II (participatory realism)" for his paper.

After my note of 28 July 2015 (Sect. 2 below), Adán responded with,

**Adánism 1:** *There is a part of your e-mail that I found deeply illuminating. It is the part ending in:*

> QBism takes the idea of irreducible randomness much further than Zeilinger or Brukner. Quantum theory advises us to recognize that no matter how we slice up nature, we will never find any pieces of it beholden to **laws** in the usual physics sense of the term. John Wheeler toyed with the idea that "the only law of nature is that there is no law," and quantum theory on a QBist reading supports this.

*THIS IS IT! I LOVE IT!*

*It may seem that you are denying the real or simply being elusive about the real, while you are actually making a strong statement about the real.*

I was very happy to see how this realization finally clicked for Adán. I hope it will the same for you, my new reader.

## 7.1  "QBism as Realism," to Steve Weinberg, 7 August 2015

**Weinbergism 1:** *When you say that you are averse to calling the Born rule a law of nature, "at least in the sense that you seem to support in some of your correspondence with David [Mermin]" I take this simply to mean that you do not take seriously the existence of any laws of nature of the sort I hope for, laws that are part of objective reality, independent of who is studying them or if anyone is studying them. I don't want to attribute to you or David or QBism any view one way or another about the laws of nature, but I have inserted a new short paragraph at the top of p. 8 in the attached version to try to clarify the issue in general terms.*

I have just spent a little while reviewing what you had written on QBism in Chap. 3 of your book[4] and in some of our correspondence. I much appreciate your effort to not "misrepresent" us, as you put it. In my last iteration to you, I didn't make a big stink of your calling QBism a form of "instrumentalism" because I felt it would be too much of a burden on you at the 11th hour. Also I knew that there would be plenty of subtleties involved in our coming to a mutual understanding of what each of us

---

[4] Steven Weinberg, *Lectures on Quantum Mechanics*, 2nd Edition, (Cambridge University Press, Cambridge, UK, 2015).

means by so simple a phrase as "laws of nature." In other words, it just didn't seem worth it to have a discussion.

In the meantime, however, I have put some effort into expressing more clearly the sense in which I see QBism as fitting into a realist project for physics. It was in response to a chart Adán Cabello of Seville has been putting together to try to capture the distinctions in various interpretations as regards their "assessment of reality." In that chart he had initially labelled QBism as "non-realist." For completeness, and so you'll understand the context of this note, I'll attach Adán's draft, titled "Map of Madness."

Below are a couple of responses I sent him, which I spent some time and care in composing. I hope you will find them clarifying too. At the end, Adán wrote me, "It may seem that you are denying the real or simply being elusive about the real, while you are actually making a strong statement about the real." That indeed is what I like to think of QBism—very much taking it out of the camp of instrumentalism—but you may well continue to disagree.

By the way, one of the pieces below includes some paragraphs that were cut from a Q & A at *Quanta Magazine*, which I see you have also participated in. I wish my photo there were so classy as yours!

## 7.2 "Realisms," to Adán Cabello, 28 July 2015

I've owed you a note for some time now. Let me finally sit down and try to devote the time to respond properly. Sorry to keep you waiting.

Realisms. The first thing I will tell you that I am *realistic* about is how little this note and your own efforts will help in setting the label game straight for QBism. For 20 years (very literally now), I have pleaded with the community to understand my efforts at understanding quantum mechanics as being part of a realist program, i.e., as an attempt to say something about what the world is like, how it is put together, and what's the stuff of it. But I have failed miserably at getting nearly anyone to understand this. You are very likely to fail too, no matter how accurate you ultimately make your chart. So many in our field—Matt Leifer, Tim Maudlin, David Albert, Avshalom Elitzur, Harvey Brown, Adrian Kent, Terry Rudolph, Travis Norsen, Sean Carroll, Paul Davies, Lee Smolin, Lucien Hardy, are a small sampling that come to mind—not to mention troves of science journalists, are simply unwilling to register the distinctions needed to have this discussion. These people are so stuck in a pre-conceived worldview, they cannot see the slightest bit out of it, even when approached by a nonstandard one of their own (i.e., a realist of a different flag).

That said, the only person in your present citation list who I would not call a realist is Asher Peres. Asher, in fact, took pride in calling himself alternately a positivist and an instrumentalist. Here are two instances where he labeled himself as such in print: http://arXiv.org/abs/quant-ph/9711003 and http://arXiv.org/abs/quant-ph/0310010. (You may note that he labelled me with the same terms as well. This was one of the key issues that made the writing of our joint article in *Physics Today* [18] so frustrating; every single sentence had to be a careful negotiation in language so that I could feel I wasn't selling my soul.) Asher was fully happy in thinking that the task of physics was solely in making better predictions from sense data to sense data.

Here's the way, I view my own realism. It can be taken straight from the playbook of Albert Einstein. He wrote in the "Reply to Criticisms" section of the Schilpp volume [15]:

> A few more remarks of a general nature concerning concepts and [also] concerning the insinuation that a concept—for example that of the real—is something metaphysical (and therefore to be rejected). A basic conceptual distinction, which is a necessary prerequisite of scientific and pre-scientific thinking, is the distinction between "sense-impressions" (and the recollection of such) on the one hand and mere ideas on the other. There is no such thing as a conceptual definition of this distinction (aside from circular definitions, i.e., of such as make a hidden use of the object to be defined). Nor can it be maintained that at the base of this distinction there is a type of evidence, such as underlies, for example, the distinction between red and blue. Yet, one needs this distinction in order to be able to overcome solipsism. Solution: we shall make use of this distinction unconcerned with the reproach that, in doing so, we are guilty of the metaphysical "original sin." We regard the distinction as a category which we use in order that we might the better find our way in the world of immediate sensations. The "sense" and the justification of this distinction lies simply in this achievement. But this is only a first step. We represent the sense-impressions as conditioned by an "objective" and by a "subjective" factor. For this conceptual distinction there also is no logical-philosophical justification. But if we reject it, we cannot escape solipsism. It is also the presupposition of every kind of physical thinking. Here too, the only justification lies in its usefulness. We are here concerned with "categories" or schemes of thought, the selection of which is, in principle, entirely open to us and whose qualification can only be judged by the degree to which its use contributes to making the totality of the contents of consciousness "intelligible." The above mentioned "objective factor" is the totality of such concepts and conceptual relations as are thought of as independent of experience, viz., of perceptions. So long as we move within the thus programmatically fixed sphere of thought we are thinking physically. Insofar as physical thinking justifies itself, in the more than once indicated sense, by its ability to grasp experiences intellectually, we regard it as "knowledge of the real."
>
> After what has been said, the "real" in physics is to be taken as a type of program, to which we are, however, not forced to cling a priori. ...
>
> The theoretical attitude here advocated is distinct from that of Kant only by the fact that we do not conceive of the "categories" as unalterable (conditioned by the nature of the understanding) but as (in the logical sense) free conventions. They appear to be a priori only insofar as thinking without the positing of categories and of concepts in general would be as impossible as breathing in a vacuum.

If I were to modify the wording of this any at all, I would emphasize the word "experience" in place of Einstein's "sensations" and "sense-impressions," and I would also probably give a more Darwinian-toned story about survival, etc., in place of Einstein's "we might better find our way in the world." But these are minor things. A little more serious would be that I *do* think we have a kind of direct evidence of "the real." It is in the very notion of experience itself; I'll come back to this later. What I want to emphasize at the moment is that I cannot see any way in which the program of QBism has ever contradicted what Einstein calls the program of "the real."

The only "sin" that QBism has ever committed is that it has held forthrightly and obstinately to the thought that the best understanding of quantum theory is obtained by recognizing that quantum states, quantum time-evolution maps, and the outcomes of quantum measurements all live within what Einstein calls the "subjective factor." But then it is just a *non sequitur* for anyone to say, "Well then everything in quantum theory, nay everything in physics, must live in the subjective factor. Solipsism! Non-realism! Anti-realism! Mysticism! QBists don't believe in reality! REALITY IS AN ILLUSION! (the headlines say)." This is because, if any of these cads were to take a moment to think about it, they would recognize that there is more to quantum mechanics than just three isolated terms (states, evolution, and measurement)—there's the full-blown theory that glues these notions together in a very particular way, and in a way that would have never been discovered without empirical science.

It is from this glue that QBism draws the "free conventions" Einstein speaks of. Take for example the Born Rule, expressed in the way Rüdiger and I find its content most revealing—for example, Eq. (8) in http://arXiv.org/abs/1003.5209,

$$Q(D_j) = (d + 1) \sum_{i=1}^{d^2} P(H_i)P(D_j|H_i) - 1 .$$

We call this a normative rule: It is a rule that any agent should strive for in his probability assignments with regard to a system he judges to have dimension $d$. It doesn't set or fix the values of the probabilities—those depend upon the agent's prior personal experiences, his computational powers, and a good amount of guesswork—but the equation itself is something he should *strive* to satisfy with his gambling commitments (i.e., probability assignments). In that sense, the Born Rule is much larger, less agent-specific than any probability assignments themselves. It's a rule that any agent should pick up and use. To that extent, it lives in what Einstein calls the "objective factor"—it lives at the level of the impersonal. And because of that, the Born Rule correlates with something that one might want to call "real," as justified only (Einstein) by the way we might use it to "better find our way in the world."[5]

So, that's one direction in which QBism points to realism. But there are other directions, and these have more to do with some of the things possibly nearer to your heart (and what you had wanted to use as a distinguishing factor between QBism

---

[5]It is worth noting that *in this aspect* at least, QBism bears a certain resemblance to structural realism. See, for instance, http://plato.stanford.edu/entries/structural-realism/. Imagine our universe at a time (if there ever was one) when there were no agents about to use the laws of probability theory as an aid in their gambles—i.e., no such agents had yet arrived out of the Darwinian goo. Were there any quantum states in the universe then? A QBist would say NO. It's not a matter of the quantum state of the universe waiting until a qualified Ph.D. student came along before having its first collapse, as John Bell joked, but that there simply weren't any quantum states. Indeed, on earth there weren't any quantum states until 1926 when Erwin Schrödinger wrote the first one down. The reason is simple: The universe is made of something else than quantum states. But then, what of the Born Rule? To this, in contrast, a QBist would say, "Aha, now there's a sensible question." For the Born Rule is among the set of relations an agent should strive to attain in his larger mesh of probability assignments. That normative rule is still lying about even when there are no agents to make use of it. As Craig Callender once paraphrased it back to me in a conversation, it's the normative rule which is nature's whisper, not the specific terms within it.

and Wheeler-Zeilinger-Brukner). Go back to the Eq. (8) advertised above. An easy consequence of this normative rule is that only in very limited circumstances can one make sharp (i.e., delta-function) probability assignments. Another way to say this is that quantum theory itself advises us that we should not think we can always have certainties. This, I believe, is what you're wanting to call "irreducible randomness," but it's nuanced in ways that I think Brukner and Zeilinger might not take note of. On the one hand, our Eq. (8) immediately gives something like what Rob Spekkens calls an "epistemic restriction" and Zeilinger calls an "information principle," but one has to remember that QBism is concerned *not* with "epistemic probabilities" (i.e., as only statements of ignorance or how much is known) or with some kind of spiritual substance in the way that Zeilinger often speaks (for instance when he says "the world is *made* of information"), but with probabilities as guides to action—de Finetti and Ramsey's notion of probability. In this way, QBist probabilities have no direct connection with what nature "must do," not even when $p = 1$. And that is a hugely important distinction between QBism and any other view of quantum theory. David Mermin stated this point very clearly in http://arXiv.org/abs/1409.2454:

> A very important difference of QBism, not only from Copenhagen, but from virtually all other ways of looking at science, is the meaning of probability 1 (or 0). In Copenhagen quantum mechanics, an outcome that has probability 1 is enforced by an objective mechanism. This was most succinctly put by Einstein, Podolsky and Rosen, though they were, notoriously, no fans of Copenhagen. Probability-1 judgments, they held, were backed up by "elements of physical reality".
>
> Bohr held that the mistake of EPR lay in an "essential ambiguity" in their phrase "without in any way disturbing". For a QBist, their mistake is much simpler than that: probability-1 assignments, like more general probability-$p$ assignments are personal expressions of a willingness to place or accept bets, constrained only by the requirement that they should not lead to certain loss in any single event. It is wrong to assert that probability assignments must be backed up by objective facts on the ground, even when $p = 1$. An expectation is assigned probability 1 if it is held as strongly as possible. Probability-1 measures the intensity of a belief: supreme confidence. It does not imply the existence of a deterministic mechanism.
>
> We are all used to the fact that with the advent of quantum mechanics, determinism disappeared from physics. Does it make sense for us to qualify this in a footnote: "Except when quantum mechanics assigns probability 1 to an outcome"?

This is what Rüdiger meant in his note to you when he said, "QBism thus takes the idea of irreducible randomness much further than Zeilinger or Brukner."[6] The way I might put it is that quantum theory advises us to recognize that no matter how we slice up nature, we will never find any pieces of it beholden to *laws* in the usual

---

[6]Rüdiger Schack has furthermore found a very fine way of putting this point in some recent talks delivered in Stellenbosch and Siegen. First he quotes Einstein, in a letter to F. Reiche and his wife, dated 15 August 1942. Einstein wrote, "I still do not believe that the Lord God plays dice. If he had wanted to do this, then he would have done it quite thoroughly and not stopped with a plan for his gambling: In for a penny, in for a pound. Then we wouldn't have to search for laws at all." Then, on the next slide Rüdiger writes, "The usual reading: Einstein advocates deterministic laws. QBist reading: There are indeed no laws. *God has done it thoroughly.* There are no laws of nature, not even stochastic ones. The world does not evolve according to a mechanism. What God has provided, on the other hand, is tools for agents to navigate the world, to survive in the world."

physics sense of the term. John Wheeler toyed with the idea that "the only law of nature is that there is no law," and quantum theory on a QBist reading supports this.

So, what backs up this position? The answer is: The general structure of quantum theory—i.e., the part correlative with "the real" (in the Einstein sense of a "free convention")—in particular the part of it you have explored so thoroughly, the Kochen-Specker and Bell theorems in all their great variety. Thus, part of your chart is definitely wrong in how it attempts to draw a distinction between QBism and Wheeler-Zeilinger-Brukner. QBism agrees with Wheeler that quantum measurements gives rise to new creation within the universe—he likened these creations to the big bang itself, see our http://arXiv.org/abs/1412.4209—but QBism maybe even goes further than Wheeler would have, I don't know, in that it holds fast to this statement even when one has a $p = 1$ prediction beforehand. Now, going that far—i.e., asserting no pre-existent properties even when $p = 1$—is indeed a point of distinction between Brukner, Zeilinger, and Kofler and QBism. You can find instances of this statement in several places in their writings (what the philosophers call the "eigenstate-eigenvalue link") [8, 35]. Another place where your table is off the mark connects to what I alluded to above when I said I have a bigger disagreement with the Einstein quote. You cite a 2005 Zeilinger paper [115] for saying, "this randomness of the individual event is the strongest indication we have of a reality 'out there' existing independently of us," but you could just as well have cited my 2002 paper http://arXiv.org/abs/quant-ph/0204146:

I would say all our evidence for the reality of the world comes from without us, i.e., not from within us. We do not hold evidence for an independent world by holding some kind of transcendental knowledge. Nor do we hold it from the practical and technological successes of our past and present conceptions of the world's essence. It is just the opposite. We believe in a world external to ourselves precisely because we find ourselves getting unpredictable kicks (from the world) all the time. If we could predict everything to the final T as Laplace had wanted us to, it seems to me, we might as well be living a dream.

To maybe put it in an overly poetic and not completely accurate way, the reality of the world is not in what we capture with our theories, but rather in all the stuff we don't. To make this concrete, take quantum mechanics and consider setting up all the equipment necessary to prepare a system in a state $\Pi$ and to measure some noncommuting observable $H$. (In a sense, all that equipment is just an extension of ourselves and not so very different in character from a prosthetic hand.) Which eigenstate of $H$ we will end up getting as our outcome, we cannot say. We can draw up some subjective probabilities for the occurrence of the various possibilities, but that's as far as we can go. (Or at least that's what quantum mechanics tells us.) Thus, I would say, in such a quantum measurement we touch the reality of the world in the most essential of ways.

However, I don't like using the term "intrinsic randomness" for all this, and you detected as much in my interview with Schlosshauer [26]. That's because "intrinsic randomness" strikes me as such a lifeless phrase in comparison to what I think is the deeper issue—namely, that all the components/pieces/slices of the world have a genuine autonomy unto themselves. And when those pieces get together, they go on to create even more autonomous stuff. This way of speaking tries to capture that there's a creativity or novelty in the world in a way that the phrase "intrinsic randomness" leaves limp. Here's how I put it in an introduction to a lecture I gave at Caltech in 2004 (reprinted in http://arXiv.org/abs/1405.2390):

A lecturer faces a dilemma when teaching a course at a farsighted summer school like this one. This is because, when it comes to research, there is often a fine line between what one thinks and what is demonstrable fact. More than that, conveying to the students what one thinks—in other words, one's hopes, one's desires, the potentest of one's premises—can be just as empowering to the students' research lives (even if the ideas are not quite right) as the bare tabulation of any amount of demonstrable fact. So I want to use one percent of this lecture to tell you what I think—the potentest of all my premises—and use the remaining ninety-nine to tell you about the mathematical structure from which that premise arises.

I think the greatest lesson quantum theory holds for us is that when two pieces of the world come together, they give birth. [Bring two fists together and then open them to imply an explosion.] They give birth to FACTS in a way not so unlike the romantic notion of parenthood: that a child is more than the sum total of her parents, an entity unto herself with untold potential for reshaping the world. Add a new piece to a puzzle—not to its beginning or end or edges, but somewhere deep in its middle—and all the extant pieces must be rejiggled or recut to make a new, but different, whole.[7]

Imagine along with Wheeler that the universe can somehow be identified with a formal mathematical system, with the universe's life somehow captured by all the decidable propositions within the system. Wheeler's "crazy" idea seems to be this. Every time an act of observer-participancy occurs (every time a quantum measurement occurs), one of the *un*decidable propositions consistent with the system is upgraded to the status of a new axiom with truth value either TRUE or FALSE. In this way, the life of the universe as a whole takes on a deeply new character with the outcome of each quantum measurement. The "intrinsic randomness" dictated by quantum theory is not so much like the flicker of a firefly in the fabric of night, but a rearrangement of the whole meaning of the universe. That is the great lesson.

But quantum mechanics is only a glimpse into this profound feature of nature; it is only a part of the story. For its focus is exclusively upon a very special case of this phenomenon: The case where one piece of the world is a highly-developed decision-making agent—an experimentalist—and the other piece is some fraction of the world that captures his attention or interest.

When an experimentalist reaches out and touches a quantum system—the process usually called quantum 'measurement'—that process gives rise to a birth. It gives rise to a little act of creation. And it is how those births or acts of creation impact the agent's expectations for other such births that is the subject matter of quantum theory. That is to say, quantum theory is a calculus for aiding us in our decisions and adjusting our expectations in a QUANTUM WORLD. Ultimately, as physicists, it is the quantum world for which we would like to say as much as we can, but that is not our starting point. Quantum theory rests at a level higher than that.

To put it starkly, quantum theory is just the start of our adventure. The quantum world is still ahead of us. So let us learn about quantum theory.

Afterward, again in http://arXiv.org/abs/1405.2390, you'll find a lot of discussion about how the word "FACTS" is not quite right for these little moments of creation (that word itself is a bit too lifeless for what I am hoping to convey)—look up the words QBoom and QBlooey:

---

[7]It is a bit of a stretch, but I have found a *wildly-speculative* idea in some recently unearthed notes from a 1974 notebook of John Wheeler's which is mildly evocative of the metaphor just given. See https://jawarchive.files.wordpress.com/2012/03/twa-1974.pdf and https://jawarchive.files.wordpress.com/2012/03/tarski.pdf, typed transcripts of which may be found in the Appendix. Despite the dubious connection to anything firmly a part of QBism, I report Wheeler's idea because it seems to me that it conveys some imaginative sense of how the notion of "birth" described here carries a very different flavor from the "intrinsic randomness" that Adán and others seem to be talking about.

- **QBoom** – cf. *kaboom*; the sought-for deanthropocentrized distillate of quantum measurement that QBism imagines powering the world. William James called it "pure experience," where "new being comes in local spots and patches which add themselves or stay away at random, independently of the rest." John Wheeler asked, "Is the entirety of existence, rather than built on particles or fields or multidimensional geometry, built on billions upon billions of elementary quantum phenomena, those elementary acts of 'observer-participancy'? ... Is what took place at the big bang the consequence of billions upon billions of these elementary 'acts of observer-participancy'?" In place of a Big Bang, the QBist wonders whether it might not be myriads and myriads of little QBooms!

- **QBlooey** – cf. *kablooey*; a QBist slur on the usual conception of the Big Bang, where the universe had its one and only creative moment at the very beginning.

but I hope this is enough to give you at least a hint of why I don't like the flavor of the term "intrinsic randomness" even if there is a technical sense in which it is correct to QBism.

I thought a little about why you might have thought QBism was noncommittal on this idea of "intrinsic randomness." Perhaps it was because of this slide which I presented in Buenos Aires:

### No Commitment to Ontology Here

Most of the time one sees Bayesian probabilities characterized as measures of ignorance or imperfect knowledge. But that description carries with it a metaphysical commitment that is not necessary for the personalist Bayesian.

Imperfect knowledge? It sounds like something that, at least in imagination, could be perfected, making all probabilities zero or one—one uses probabilities only because one does not know the true pre-existing state of affairs.

All that matters is that there is *uncertainty* for whatever reason. There might be uncertainty because there is ignorance of a true state of affairs, but there might be uncertainty because the world itself does not yet know what it will give—i.e., there is an objective indeterminism.

If so, then maybe this will help dispel the confusion: You should understand that the slide was only meant to make a statement about the subjective Bayesian notion of probability. The notion itself is noncommittal. However, subjective Bayesian probability *combined* with the Born Rule and Kochen-Specker considerations, etc., is very committal. In fact, I drew the words for that slide from Footnote 14 of http://arXiv.org/abs/1003.5209, but what I left out was the footnote's very ending sentence: "As will be argued in later sections, QBism finds its happiest spot in an unflinching combination of 'subjective probability' with 'objective indeterminism'." See also this abstract from another talk where I point out how QBism's metaphysics may not have been amenable to de Finetti's own:

### Something on QBism

The term QBism, invented in 2009, initially stood for Quantum Bayesianism, a view of quantum theory a few of us had been developing since 1993. Eventually, however, I. J. Good's warning that there are 46,656 varieties of Bayesianism came to bite us, with some Bayesians feeling their good name had been hijacked. David Mermin suggested that the B in QBism should more accurately stand for "Bruno," as in Bruno de Finetti, so that we would at least get the variety of (subjective) Bayesianism right. The trouble is QBism incorporates a kind of metaphysics that even Bruno de Finetti might have rejected! So, trying to be as true to our story as possible, we momentarily toyed with the idea of associating the B with what the early 20th-century United States Supreme Court Justice Oliver Wendell Holmes Jr. called "bettabilitarianism." It is the idea that the world is loose at the joints, that indeterminism plays a real role in the world. In the face of such a world, what is an active agent to do but *participate* in the uncertainty that is all around him? As Louis Menand put it, "We cannot know what consequences the universe will attach to our choices, but we can bet on them, and we do it every day." This is what QBism says quantum theory is about: How to best place bets on the consequences of our actions in this quantum world. But what an ugly, ugly word, *bettabilitarianism*! Therefore, maybe one should just think of the B as standing for no word in particular, but a deep idea instead: That the world is so wired that our actions as active agents actually *matter*. Our actions and their consequences are not eliminable epiphenomena. In this talk, I will describe QBism as it presently stands and give some indication of the many things that remain to be developed.

So, I write all these words to say, *PLEASE* don't call us "non-realist" in print. We have endured enough hardship because of that damned label. That label probably won't ever leave us (my first realism expressed above), but I would like to think that you, our friend, never contributed any of your own to the trouble.

Then, you say, "Well then, what instead should I call the category in which QBism sits? If not non-realism, what?" I have an answer. But before I propose it, I want to say something about one more *realism* of QBism. By lifting the quantum formalism from being directly descriptive of the world external to the agent and instating it instead as a normative prescription for aiding in his survival, at the same time as asserting that agents are physical systems like any others, QBism breaks into a territory the vast majority of those declaring they have a scientific worldview would be loath to enter. And that is that the agents (observers) matter as much as electrons and atoms in the construction of the actual world—the agents using quantum theory are not incidental to it. Johannes Kofler said recently to Rüdiger Schack, "I am convinced we are biological machines." I told Rüdiger his reply should have been, "The lesson of QBism is that there are *no* machines, period." In the terms that Danny Greenberger laid out in http://www.iqoqi-vienna.at/can-a-computer-ever-become-conscious/, everything in the universe has the potential to "push the big red button." And nothing should be forgotten in that regard, from the lowliest electron to the very user of quantum theory. I've always admired the way my student John DeBrota put the key point in his graduate application: "From [QBism] I have learned that the universe can be moved—that it is not a masterfully crafted mechanical automaton, but instead an unfinished book, brimming with creation and possibility, of which every agent is an author." If that too is a reality of QBism, then it should be recognized in its categorization.

It's the reality that the observer, the agent is an active and non-negligible *pariticipator* in the universe as John Wheeler emphasized so many times. Certainly, recall Wheeler's game of 20 questions, surprise version (taken from "Frontiers of Time," 1979) [69]:

> The Universe can't be Laplacean. It may be higgledy-piggledy. But have hope. Surely someday we will see the necessity of the quantum in its construction. Would you like a little story along this line?
>
> Of course! About what?
>
> About the game of twenty questions. You recall how it goes—one of the after-dinner party sent out of the living room, the others agreeing on a word, the one fated to be a questioner returning and starting his questions. "Is it a living object?" "No." "Is it here on earth?" "Yes." So the questions go from respondent to respondent around the room until at length the word emerges: victory if in twenty tries or less; otherwise, defeat.
>
> Then comes the moment when we are fourth to be sent from the room. We are locked out unbelievably long. On finally being readmitted, we find a smile on everyone's face, sign of a joke or a plot. We innocently start our questions. At first the answers come quickly. Then each question begins to take longer in the answering—strange, when the answer itself is only a simple "yes" or "no." At length, feeling hot on the trail, we ask, "Is the word 'cloud'?" "Yes," comes the reply, and everyone bursts out laughing. When we were out of the room, they explain, they had agreed not to agree in advance on any word at all. Each one around the circle could respond "yes" or "no" as he pleased to whatever question we put to him. But however he replied he had to have a word in mind compatible with his own reply—and with all the replies that went before. No wonder some of those decisions between "yes" and "no" proved so hard!
>
> And the point of your story?
>
> Compare the game in its two versions with physics in its two formulations, classical and quantum. First, we thought the word already existed "out there" as physics once thought that the position and momentum of the electron existed "out there," independent of any act of observation. Second, in actuality the information about the word was brought into being step by step through the questions we raised, as the information about the electron is brought into being, step by step, by the experiments that the observer chooses to make. Third, if we had chosen to ask different questions we would have ended up with a different word—as the experimenter would have ended up with a different story for the doings of the electron if he had measured different quantities or the same quantities in a different order. Fourth, whatever power we had in bringing the particular word "cloud" into being was partial only. A major part of the selection—unknowing selection—lay in the "yes" or "no" replies of the colleagues around the room. Similarly, the experimenter has some substantial influence on what will happen to the electron by the choice of experiments he will do on it; but he knows there is much impredictability about what any given one of his measurements will disclose. Fifth, there was a "rule of the game" that required of every participator that his choice of yes or no should be compatible with some word. Similarly, there is a consistency about the observations made in physics. One person must be able to tell another in plain language what he finds and the second person must be able to verify the observation.

In honor of that, and because I do think it captures a big part of the idea we are all expressing, I would suggest calling our category "Participatory Realism." I would say everyone you presently categorize as non-realist (with the exception of Asher) partakes to some extent in a kind of participatory realism. Wheeler has more of it than Zeilinger, and Zeilinger may have more of it than Bohr, and Bohr more than Rovelli, but they all have a bit of it. QBism carries it to its logically enforced extreme.

Finally, if QBism and the others in the right-side table are categorized as "partic-ipatory realism," the ones on the left side can't be called simply realist. For they are a very specialized species of the genus: All they really want is their hidden variable, somewhere, somehow (Spekkens too is certainly guilty of this even though he would try to deny it). And they want physicists and humankind more generally to be rele-gated to being inessential epiphenomena in the universe. They want the physicist to really have nothing to with Physics with a capital P. Thus, here are some suggestions:

> *dumb realism*
>      ("lacking intelligence or good judgment; stupid; dull-witted")
> *static realism*
> *stillborn realism*
> *base realism*
>      ("morally low; without estimable qualities; dishonorable; cowardly")

But whatever you do, *PLEASE* call QBism *"any damned thing but"* non-realist! (See this video of Johnny Cash singing "A Boy Named Sue," starting at the 2:53 mark of https://www.youtube.com/watch?v=WOHPuY88Ry4.)

## 7.3  "Slicing the Euglena's Tail," to Adán Cabello, 28 July 2015

And one further note to address this remark you made to Rüdiger in your last note:

**Adánism 2:**  *This may be the point! For QBism, quantum theory is "a 'user's man-ual' any agent can pick up." End of the story. While Zeilinger is trying to get insight about how nature works from that manual. Am I right?*

You are wrong. I told the story below in my talk in Bueno Aires, but it probably went by too fast for you to catch. See the missing paragraphs below from my inter-view with Amanda Gefter: https://www.quantamagazine.org/20150604-quantum-bayesianism-qbism/. They address your contention head on.

**Amanda:** QBism says that quantum mechanics is a "single-user theory."

**Chris:** Any of us can use quantum theory, but only for ourselves. There's a little single-celled thing called a Euglena that has a tail coming off of it. The tail arose from evolutionary pressures, so that the Euglena can move from environments where there are depleted nutrients to environments where there's an abundance of nutrients. It's a tool. Quantum mechanics is like the Euglena's tail. It's something we evolved in 1925 and since it's been shown to be such a good tool, we keep using it and we pass it on to our children. The tail is a single-user tail. But we can look at the tail and ask things like, what might we learn about the environment by studying its structure? We might notice the tail is not completely circular and that might tell us something about the viscosity of the medium it's traveling through. We might look at the ratio of the length of it to the width of it in various places and that might tell us about features of the environment. So quantum mechanics is a single-user theory, but by dissecting it, you can learn something about the world that all of us are immersed in.

**Amanda:** So eventually objectivity comes in?

**Chris:** I hope it does. Ultimately I view QBism as a quest to point to something in the world and say, that's intrinsic to the world. But I don't have a conclusive answer yet. Let's take the point of view that quantum mechanics is a user's manual. A user's manual *for me*. A philosopher will quickly say, well that's just instrumentalism. "Instrumentalism" is always prefaced by a "just." But that's jumping too quickly to a conclusion. Because you can always ask—you should always ask—what is it about the world that compels me to adopt this instrument rather than that instrument? A quantum state is a user's manual of probabilities. But how does it determine the probabilities? Well there's a little mathematical formula called the Born rule. And then you should ask, why that formula? Couldn't it have been a different formula? Yes, it might have been different. The fact that we adopt this formula rather than some other formula is telling us something about the character of the world as it is, independent of us. If we can answer the question "Why the Born Rule?" or John Wheeler's question "Why the quantum?" then we'll be making a statement about how the world is, one that's not "just" instrumentalism.

## 7.4 "Denouement," to Johannes Kofler, 6 October 2014

Here is the result that came out of your kindness of lending me your Schilpp volume and letting me subject its spine to the copier.

Let me just tell you a bit about how this little quest fits into my bigger agenda (and is something I view as a mandate for QBism). It's that I take absolutely seriously John Wheeler's "idea for an idea" (as he would say) that the "elementary quantum phenomenon" might be taken as the ultimate building block of reality. I also take absolutely seriously his idea that within every "phenomenon" is an instance of creation, not unlike what one usually exclusively associates with the Big Bang. This caused John to speculate that "perhaps the Big Bang is here all around us." (This idea that *every* quantum measurement results in an instance of creation is also connected to the QBist rejection of the EPR reality criterion. Even outcomes following probability-1 assignments are for us are instances of creation. We could not have that if probability-1 in fact meant pre-existence. But this is an aside, connected more to the note I will send you following this one.)

So getting Bohr's idea of "phenomenon" straight to myself is quite important from my perspective. I think much of what Bohr says was flawed, but that doesn't mean I still don't have several things to learn from him. What he calls "phenomenon" resembles in many ways what a set of philosophers early last century (William James, John Dewey, Ralph Barton Perry, and some others) called "pure experience" … though they explored their idea from all angles and in great detail, rather than spewing the same few phrases over and over like Bohr did. Importantly James saw his notion of "pure experience" as the ultimate building block of reality—i.e., like Wheeler of Bohr's "phenomenon." The reason Bohr still interests me, however, even though his story is hardly written at all compared to the great corpus of James and Dewey, is that he at least makes a connection to the quantum, whereas they could not. In fact, as best I can tell, when Bohr continually talks about the indivisibility of the quantum, he is using that as a synonym for what he calls in other contexts "the quantum postulate." So he saw the "phenomena" as the very reason for being for quantum mechanics.

The greatest and most difficult task will be—after developing QBism sufficiently well to get a clear view of all that is around it—to return to what Bohr tried to jump to prematurely with his "classically described agencies of observation" and *disambiguate* the "pure experience" notion from the agents gambling upon them. Only then can one imagine "pure experience" or "phenomenon" as an "ultimate building block for all reality."

**Acknowledgements** I thank Ted Hänsch and Ignacio Cirac for affording me the leisure to think on these things in the quiet of little Garching some time before writing them down. I thank Blake Stacey for advice on this manuscript.

## Appendix: Transcription from John Wheeler's Notebook

[Text taken from photographs of notebook posted at https://jawarchive.wordpress. com/.]

4–6 February 1974 <u>draft notes</u> for discussion with Dana Scott; also with Simon Kochen, Charles Patton and Roger Penrose.

### ADD "PARTICIPANT" TO "UNDECIDABLE PROPOSITIONS" TO ARRIVE AT PHYSICS

John A. Wheeler

#### Brief

We consider the quantum principle. Of all of the well-analyzed features of physics there is none more central, and none of an origin more mysterious. This note identifies its key idea as <u>participation</u>; and its point of origin, as the "undecidable propositions" of mathematical logic. On this view physics is not machinery. Logic is not oil occasionally applied to that machinery. Instead everything, physics included, derives from two parents, and is nothing but cathode-tube image of the interplay between them. One is the "participant." The other is the complex of undecidable propositions of mathematical logic. The "participator" assigns true-false values to appropriate ones among these propositions at his own free will. As he does so, the corresponding world unrolls on his screen. No participator, no world!

#### Comments

1. <u>The quantum principle and physics.</u>

The quantum principle is taken here to be the pervasive unifying element of all of physics. It would be a favorable sign to find the quantum principle derivable from mathematical logic along the foregoing lines; and to find the opposite would be a decisive blow against these views.

2. Start with what formal system?

Take a formal system. Enlarge it to a new formal system, and that again to a new formal system, and so on, by resolving undecidable propositions ("act of participation"). Will the system become so complex that it can and must be treated by statistical means? Will such a treatment make it irrelevant, or largely irrelevant, with what particular formal system one started?

3. Lack of commutativity

After the system has grown to a certain level of complexity, one can imagine a difference in the subsequent development, according as decisions about appropriate "undecidable propositions" are taken in the order AB or the order BA. One might focus on this point in trying to locate something like the quantum principle as already contained in mathematical logic.

4. "Reality"

The propositions are not propositions about anything. They are the abstract building blocks, or "pregeometry," out of which "reality" is conceived as being built.

Mon 25 Feb '74  PATTON, FOLLOW-UP OF SCOTT DISCUSSION

Can we find "Tr" in our theory and A in our theory such that for all provable statements it comes out OK and yet also to unprov. statements assign truth values[?] Tarski says can't do; have to have a bigger theory; have to have someone on outside imposing what's true & what's false. Truth is thus a "meta" concept. "Participator" here! Logic can't live by itself. No wonder Boolean system won't fly.

# References

1. Appleby, D.M.: The Bell-Kochen-Specker theorem. Stud. Hist. Philos. Sci. **36**, 1–28 (2005)
2. Appleby, D.M.: Facts, values and quanta. Found. Phys. **35**, 627–668 (2005)
3. Appleby, D.M.: Probabilities are single-case, or nothing. Opt. Spectro. **99**, 447–456 (2005)
4. Appleby, D.M.: Concerning dice and divinity. In: Adenier, G. Fuchs, C.A., Khrennikov, A. Yu. (eds.) Foundations of Probability and Physics—4, pp. 30–39. AIP Conference Proceedings, vol. 889, American Institute of Physics, Melville, NY. arXiv:quant-ph/0611261 (2007)
5. Appleby, D.M., Ericsson, Å., Fuchs, C.A.: Properties of QBist state spaces. Found. Phys. **41**, 564–579 (2010)
6. Appleby, D.M.: Mind and matter: a critique of Cartesian thinking. In: Atmanspacher, H., Fuchs, C.A (eds.) The Pauli-Jung conjecture and its impact today, pp. 7–36. Imprint Academic, Exeter, UK. arXiv:1305.7381v1 (2013)
7. Appleby, D.M, Fuchs, C.A., Stacey, B.C., Zhu, H.: The Qplex: A Novel Arena for Reconstructing Quantum Theory. Forthcoming (2016)

8. Brukner, Č., Zeilinger, A.: Conceptual inadequacy of the Shannon information in quantum measurements. Phys. Rev. A **63**, 022113 (2001)
9. Bub, J.: Bananaworld: Quantum Mechanics for Primates. Oxford University Press, Oxford, p. 232 (2016)
10. Cabello, A.: Interpretations of Quantum Theory: A Map of Madness. arXiv:1509.04711 [quant-ph] (2015)
11. Caves, C.M., Fuchs, C.A., Schack, R.: Quantum probabilities as Bayesian probabilities. Phys. Rev. A **65**, 022305 (2002)
12. Caves, C.M., Fuchs, C.A., Schack, R.: Unknown quantum states: the quantum de Finetti representation. J. Math. Phys. **43**, 4537–4559 (2002)
13. Caves, C.M., Fuchs, C.A., Schack, R.: Conditions for compatibility of quantum-state assignments. Phys. Rev. A **66**, 062111 (2002)
14. Caves, C.M., Fuchs, C.A., Schack, R.: Subjective probability and quantum certainty. Stud. Hist. Philos. Sci. **38**, 255–274 (2007)
15. Einstein, A.: Remarks concerning the essays brought together in this co-operative volume. In: Schilpp, P.A. (ed.) Albert Einstein: Philosopher-Scientist, pp. 665–688. Tudor Publishing Co., New York (1949)
16. Fuchs, C.A.: Quantum mechanics as quantum information (and only a little more). In: Khrennikov, A. (ed.) Quantum Theory: Reconsideration of Foundations, pp. 463–543. Växjö University Press, Växjö, Sweden. arXiv:quant-ph/0205039 (2002)
17. Fuchs, C.A.: The Anti-Växjö interpretation of quantum mechanics. In: Krennikov, A. (ed.) Quantum Theory: Reconsideration of Foundations, pp. 99–116. Växjö University Press, Växjö, Sweden. arXiv:quant-ph/0204146 (2002)
18. Fuchs, C.A., Peres, A.: Quantum theory needs no 'interpretation'. Phys. Today **53**(3), 70–71 (2000)
19. Fuchs, C.A., Peres, A.: Quantum theory—interpretation, formulation, inspiration: fuchs and peres reply. Phys. Today **53**(9), 14, 90 (2000)
20. Fuchs, C.A., Schack, R., Scudo, P.F.: A de Finetti representation theorem for quantum process tomography. Phys. Rev. A **69**, 062305 (2004)
21. Fuchs, C.A., Schack, R.: Unknown quantum states and operations, a Bayesian view. In: Paris, M.G.A., Reháček, J. (eds.) Quantum Estimation Theory, pp. 151–190. Springer, Berlin arXiv:quant-ph/0404156 (2004)
22. Fuchs, C.A.: Delirium quantum: or, where i will take quantum mechanics if it will let me. In: Adenier, A., Fuchs, C.A., Khrennikov, A.Yu. (eds.) Foundations of Probability and Physics—4, pp. 438–462. AIP Conference Proceedings Vol. 889, American Institute of Physics, Melville, NY arXiv:0906.1968 [quant-ph] (2007)
23. Fuchs, C.A., Schack, R.: Priors in quantum bayesian inference. In: Accardi, L. et al (eds.) Foundations of Probability and Physics—5, pp. 255–259. AIP Conference Proceedings, vol. 1101, (American Institute of Physics, Melville, NY. arXiv:0906.1714 [quant-ph] (2009)
24. Fuchs, C.A.: QBism, the Perimeter of Quantum Bayesianism. arXiv:1003.5209 [quant- ph] (2010)
25. Fuchs, C.A.: Coming of Age with Quantum Information: Notes on a Paulian Idea. Cambridge University Press, Cambridge, UK (2010)
26. Fuchs, C.A.: Interview with a quantum Bayesian. In: Schlosshauer, M. (ed.) Elegance and Enigma: The Quantum Interviews. Springer, Berlin, Frontiers Collection. arXiv:1207.2141 [quant-ph] (2011)
27. Fuchs, C.A., Schack, R.: A quantum-Bayesian route to quantum-state space. Found. Phys. **41**, 345–356 (2011)
28. Fuchs, C.A., Schack, R.: Bayesian conditioning, the reflection principle, and quantum decoherence. In: Ben-Menahem, Y., Hemmo, M. (eds.) Probability in Physics, pp. 233–247. Springer, Berlin, Frontiers Collection (2012)
29. Fuchs, C.A., Schack, R.: Quantum-Bayesian coherence. Rev. Mod. Phys. **85**, 1693–1715 (2013)

30. Fuchs, C.A. My struggles with the block universe: selected correspondence, January 2001—May 2011, Stacey, B.C. (ed.), Schlosshauer, M. (foreword). arXiv:1405.2390 [quant-ph] (2014)
31. Fuchs, C.A., Schack, R.: Quantum measurement and the Paulian idea. In: Atmanspacher, A., Fuchs, C.A. (eds.) The Pauli-Jung Conjecture and Its Impact Today, pp. 93–107. Imprint Academic, Exeter, UK (2014)
32. Fuchs, C.A., Mermin, N.D., Schack, R.: An introduction to QBism with an application to the locality of quantum mechanics. Am. J. Phys. **82**, 749–754 (2014)
33. Fuchs, C.A., Mermin, N.D., Schack, R.: Reading QBism: reply to Nauenberg. Am. J. Phys. **83**, 198 (2015)
34. Fuchs, C.A., Schack, R.: QBism and the Greeks: why a quantum state does not represent an element of physical reality. Physica Scripta **90**, 015104 (2015)
35. Kofler, J., Zeilinger, A.: Quantum information and randomness. Eur. Rev. **18**, 469–480 (2010)
36. Ladyman, J., Ross, D., Spurrett, D., Collier, J.: Everything Must Go: Metaphysics Naturalized, p. 184. Oxford University Press, Oxford (2007)
37. Mermin, N.D.: What's bad about this habit. Phys. Today **62**(5), 8–9 (2009); reprinted in reprinted in Mermin, N.D.: Why Quark Rhymes with Pork and other Scientific Diversions, Chapter 33. Cambridge University Press, Cambridge, UK (2016)
38. Mermin, N.D.: Mermin habitually answers opinions, real and abstract: Mermin replies. Phys. Today **62**(9), 14–15 (2009)
39. Mermin, N.D.: Quantum mechanics: fixing the shifty split. Phys. Today **65**(7), 8–10 (2012); reprinted in Mermin, N.D.: Why Quark Rhymes with pork and other Scientific Diversions, Chapter 31. Cambridge University Press, Cambridge, UK (2016)
40. Mermin, N.D.: Measured responses to quantum Bayesianism: Mermin replies **65**(12), 12–15 (2012)
41. Mermin, N.D.: Annotated Interview with a QBist in the Making. arXiv:1301.6551 [quant-ph] (2013)
42. Mermin, N.D.: Impressionism, realism, and the aging of Ashcroft and Mermin: reply to Menéndez. Phys. Today **66**(7), 8 (2013)
43. Mermin, N.D.: What I think about now. Phys. Today **67**(3), 8–9 (2014); reprinted in Mermin, N. D.: Why Quark Rhymes with Pork and other Scientific Diversions, Chapter 32. Cambridge University Press, Cambridge, UK (2016)
44. Mermin, N.D.: Classical and quantum framing of the Now: Mermin replies. Phys. Today **67**(9), 8–9 (2014)
45. Mermin, N.D.: QBism puts the scientist back into science. Nature **507**, 421–423 (2014)
46. Mermin, N.D.: QBism in the New Scientist. arXiv:1406.1573 [quant-ph] (2014)
47. Mermin, N.D.: Why QBism is Not the Copenhagen Interpretation and What John Bell Might Have Thought of It. arXiv:1409.2454 [quant-ph] (2014); reprinted in Mermin, N. D.: Why Quark Rhymes with Pork and other Scientific Diversions, Chapter 32. Cambridge University Press, Cambridge, UK (2016)
48. Penrose, R.: The Emperor's New Mind: Concerning Computers, Minds, and the Laws of Physics, p. 295. Oxford University Press, Oxford (1990)
49. Schack, R., Brun, T.A., Caves, C.M.: Quantum Bayes rule. Phys. Rev. A **64**, 014305 (2001)
50. Stacey, B.C.: SIC-POVMs and Compatibility Among Quantum States. Mathematics **4**, 36 arXiv:1404.3774 [quant-ph] (2016)
51. Stacey, B.C.: Von Neumann Was Not a Quantum Bayesian. Phil. Trans. Roy. Soc. A **374**, 20150235. arxiv:1412.2409 [physics. hist-ph] (2014)
52. Stacey, B.C.: Multiscale structure in eco-evolutionary dynamics, Chapters 5 and 10. Ph.D. thesis, Brandeis University. arXiv:1509.02958 [q-bio.PE] (2015)
53. Timpson, C.G.: Quantum Bayesianism: a study. Stud. Hist. Philos. Sci. **39**, 579–609 (2008)
54. von Baeyer, H.C.: QBism: The Future of Quantum Physics. Harvard University Press, Cambridge, MA (2016)

55. Wheeler, J.A.: Transcending the law of conservation of leptons. In: Atti del Convegno Internazionale sul Tema: The Astrophysical Aspects of the Weak Interaction. Cortona "Il Palazzone,", pp. 133–164. 10–12 Giugno (1970), Accademia Nazionale die Lincei, Quaderno N. **157** (1971)

56. Wheeler, J.A.: From relativity to mutability. In: Mehra, J., Reidel, D. (eds.) The Physicist's Conception of Nature, Dordrecht, pp. 202–247 (1973)

57. Wheeler, J.A.: From Mendeléev's atom to the collapsing star. In: Seeger, R.J., Cohen, R.S. (eds.) Philosophical Foundations of Science, pp. 275–301. Reidel, Dordrecht (1974)

58. Wheeler, J.A.: The universe as home for man. Am. Sci. **62**, 683–691 (1974). This is an abridged, early version of Ref. [61]

59. Wheeler, J.A.: From magnetic collapse to gravitational collapse: levels of understanding magnetism. In: Canuto, V. (ed.) Role of Magnetic Fields in Physics and Astrophysics, pp. 189–221. Annals of the New York Academy of Sciences **257**, NY Academy of Sciences, NY (1975)

60. Wheeler, J.A.: Another big bang? Am. Sci. **63**, 138 (1975). This is a letter in reply to a reader's comments on Ref. [58]

61. Wheeler, J.A.: The universe as home for man. In: Gingerich, O. (ed.) The Nature of Scientific Discovery: A Symposium Commemorating the 500th Anniversary of the Birth of Nicolaus Copernicus, pp. 251–296, discussion pp. 575–587. Smithsonian Institution Press, City of Washington (1975)

62. Wheeler, J.A., Patton, C.M.: Is physics legislated by cosmogony? In: Isham, C.J., Isham, R., Penrose, R., Sciama, D.M. (eds.) Quantum Gravity: An Oxford Symposium, pp. 538–605. Clarendon Press, Oxford (1975)

63. Wheeler, J.A.: Include the observer in the wave function? Fundamenta Scientiae: Seminaire sur les Fondements des Sciences (Strasbourg) **25**, 9–35 (1976)

64. Wheeler, J.A., Patton, C.M.: Is physics legislated by cosmogony? In: Duncan, R., Weston-Smith, M. (eds.) Encyclopedia of ignorance: everything you ever wanted to know about the unknown, pp. 19–35. Pergamon, Oxford, (1977). This is an abridged version of Ref. [62]

65. Wheeler, J.A.: Genesis and observership. In: Butts, R.E., Hintikka, J. (eds.) Foundational Problems in the Special Sciences: Part Two of the Proceedings of the Fifth International Congress of Logic, Methodology and Philosophy of Science, London, Canada, 1975, pp. 3–33. D. Riedel, Dordrecht (1977)

66. Wheeler, J.A.: Include the observer in the wave function? In: Leite Lopes, J., Paty, M. (eds.) Quantum Mechanics, a Half Century Later: Papers of a Colloquium on Fifty Years of Quantum Mechanics, Held at the University Louis Pasteur, Strasbourg, May 2–4, 1974, pp. 1–18. D. Reidel, Dordrecht (1977). This is a reprint of Ref. [63]

67. Wheeler, J.A.: The 'past' and the 'delayed-choice' double-slit experiment. In: Marlow, A.R. (ed.) Mathematical Foundations of Quantum Theory, pp. 9–48. Academic Press, New York (1978)

68. Wheeler, J.A.: Parapsychology—a correction. Science **205**, 144 (1979). This correction refers to something stated in Wheeler's talk at the annual meeting of the AAAS (reported later in corrected form in Ref. [77])

69. Wheeler, J.A.: Frontiers of time. In: Toraldo di Francia, G. (ed.) Problems in the Foundations of Physics, Proceedings of the International School of Physics "Enrico Fermi," Course LXXII, pp. 395–492. North-Holland, Amsterdam (1979)

70. Wheeler, J.A.: The quantum and the universe. In: de Finis, F. (ed.) Relativity, Quanta, and Cosmology in the Development of the Scientific Thought of Albert Einstein, vol. II, pp. 807–825. Johnson Reprint Corp, New York (1979)

71. Wheeler, J.A.: The Superluminal, p. 68. New York Review of Books, 27 September (1979)

72. Wheeler, J.A.: Collapse and quantum as lock and key. Bull. Am. Phys. Soc. Ser. II **24**, 652–653 (1979)

73. Wheeler, J.A.: Beyond the black hole. In: Woolf, H. (ed.) Some Strangeness in the Proportion: A Centennial Symposium to Celebrate the Achievements of Albert Einstein, pp. 341–375, discussion pp. 381–386. Addison-Wesley, Reading, MA (1980)

74. Wheeler, J.A.: Pregeometry: motivations and prospects. In: Marlow, A.R. (ed.) Quantum Theory and Gravitation: Proceedings of a Symposium Held at Loyola University, New Orleans, May 23–26, 1979, pp. 1–11. Academic Press, New York (1980)
75. Wheeler, J.A.: Law without law. In: Medawar, P., Shelley, J.H. (eds.) Structure in Science and Art, pp. 132–154. Elsevier, Amsterdam (1980)
76. Wheeler, J.A.: Delayed-choice experiments and the Bohr-Einstein dialogue. American Philosophical Society and the Royal Society: Papers Read at a Meeting. June 5, 1980, pp. 9–40. American Philosophical Society, Philadelphia (1980)
77. Wheeler, J.A.: Not consciousness but the distinction between the probe and the probed as central to the elemental quantum act of observation. In: Jahn, R.G. (ed.) The Role of Consciousness in the Physical World, pp. 87–111. Westview Press, Boulder, CO (1981)
78. Wheeler, J.A.: The participatory universe. Science 81(2)(5), 66–67 (1981)
79. Wheeler, J.A.: The elementary quantum act as higgledy-piggledy building mechanism. In: Castell, L., von Weizsäcker (eds.) Quantum Theory and the Structures of Time and Space: Papers Presented at a Conference Held in Tutzing, July, 1980, pp. 27–30. Carl Hanser, Munich (1981)
80. Wheeler, J.A.: The computer and the universe. Int. J. Theor. Phys. 21, 557–572 (1982)
81. Wheeler, J.A.: Particles and geometry. In: Breitenlohner, P., Dürr, H. P. (eds.) Unified theories of elementary particles: critical assessment and prospects, pp. 189–217. Lecture Notes in Physics, vol. 160, Springer, Berlin (1982)
82. Wheeler, J.A.: Bohr, Einstein, and the strange lesson of the quantum. In: Elvee, R.Q. (ed.) Mind in Nature: Nobel Conference XVII, Gustavus Adolphus College, St. Peter, Minnesota, pp. 1–23, and discussions pp. 23–30, 88–89, 112–113, and 148–149. Harper & Row, San Francisco, CA (1982)
83. Wheeler, J.A.: Physics and Austerity: Law without Law. University of Texas preprint, 1–87 (1982)
84. Wheeler, J.A.: Black holes and new physics. Discovery: Research and Scholarship at the University of Texas at Austin 7(2), 4–7 (Winter 1982)
85. Wheeler, J.A.: On recognizing 'law without law': oersted medal response at the joint APS-AAPT meeting, New York, 25 January 1983. Am. J. Phys. 51, 398–404 (1983)
86. Wheeler, J.A.: Elementary quantum phenomenon as building unit. In: Meystre, P., Scully, M.O. (eds.) Quantum Optics. Experimental gravity, and measurement theory, pp. 141–143. Plenum Press, New York (1983)
87. Wheeler, J.A.: Law without law. In: Wheeler, J.A., Zurek, W.H. (eds.) Quantum Theory and Measurement, pp. 182–213. Princeton University Press, Princeton (1983)
88. Wheeler, J.A.: Guest editorial: the universe as home for man. In: Snow, T.P. (ed.) The Dynamic Universe: An Introduction to Astronomy, pp. 108–109. West Pub. Co., St. Paul, Minnesota, (1983). This is an excerpt from Ref. [61]
89. Wheeler, J.A.: Physics and austerity. In: Masculescu, I. (ed.) Krisis, pp. 671–675, vol. 1, No. 2, Klinckscieck, Paris (1983)
90. Wheeler, J.A.: Quantum gravity: the question of measurement. In: Christensen, S.M. (ed.) Quantum Theory of Gravity: Essays in Honor of the 60th Birthday of Bryce S. DeWitt, pp. 224–233. Adam Hilger, Bristol (1984)
91. Wheeler, J.A.: Bits, quanta, meaning. In: Giovannini, A., Mancini, F., Marinaro, M. (eds.) Problems in Theoretical Physics pp. 121–141. University of Salerno Press, Salerno (1984)
92. Wheeler, J.A.: Bits, Quanta, Meaning. Theoretical Physics Meeting: Atti del Convegno. Amalfi, 6–7 Maggio 1983, pp. 121–134. Edizioni Scientifiche Italiene, Naples (1984)
93. Wheeler, J.A.: Bohr's 'phenomenon' and 'law without law'. In: Casati, G. (ed.) Chaotic Behavior in Quantum Systems: Theory and Applications, pp. 363–378. Plenum Press, New York (1985)
94. Wheeler, J.A.: Niels Bohr, the Man. Phys. Today 38(10), 66–72 (1985)
95. Wheeler, J.A.: Delayed-choice experiments and the Bohr-Einstein dialogue. In: Mitra, A.N., Kothari, L.S., Singh, V., Trehan, S.K. (eds.) Niels Bohr: A Profile, pp. 139–168. Indian National Science Academy, New Delhi (1985). This is a reprint of Ref. [76]

96. Wheeler, J.A.: Hermann Weyl and the unity of knowledge. Am. Sci. **74**, 366–375 (1986)
97. Wheeler, J.A.: Niels Bohr: The Man and his legacy. In: de Boer, J., Dal, E., Ulfbeck, O. (eds.) The Lesson of Quantum Theory, pp. 355–367. Elsevier, Amsterdam (1986)
98. Wheeler, J.A.: 'Physics as meaning circuit': three problems. In: Moore, G.T., Scully, M.O. (eds.) Frontiers of Nonequilibrium Statistical Physics, pp. 25–32. Plenum Press, New York (1986)
99. Wheeler, J.A.: Foreword. In: Barrow, J.D., Tipler, F.J. (eds.) The Anthropic Cosmological Principle, pp. vii–ix. Oxford University Press, Oxford (1986)
100. Wheeler, J.A.: How come the quantum. In: Greenberger, D.M. New Techniques and Ideas in Quantum Measurement Theory, pp. 304–316. Ann. New York Acad. Sci. **480** (1987)
101. Wheeler, J.A.: Foreword. In: Weyl, H., Pollard, S., Bole, T. (trans.) The Continuum: A Critical Examination of the Foundation of Analysis, pp. ix–xiii. Thomas Jefferson University Press, Kirksville, MO (1987). This is an excerpt from Ref. [96]
102. Wheeler, J.A.: World as system self-synthesized by quantum networking. IBM J. Res. Dev. **32**, 4–15 (1988)
103. Wheeler, J.A.: World as system self-synthesized by quantum networking. In: Agazzi, E. (eds) Probability in the Sciences, pp. 103–129. Kluwer, Dordrecht (1988). This is a reprint of Ref. [102]
104. Wheeler, J. A.: Hermann Weyl and the unity of knowledge. In: Deppert, W. et al (ed.) Exact Sciences and their Philosophical Foundations, pp. 366–375. Lang, Frankfurt am Main (1988) This is an expanded version of Ref. [96]
105. Wheeler, J.A.: Bits, quanta, meaning. In: Giovannini, A., Mancini, F., Marinaro, M., Rimini, A. (eds.) Festschrift in Honour of Eduardo R. Caianiello, pp. 133–154. World Scientific, Singapore (1989)
106. Wheeler, J.A.: Information, Physics, Quantum: the Search for Links. In: Kobayashi, S., Ezawa, H., Murayama, Y., Nomura, S. (eds.) Proceedings of the 3rd International Symposium on Foundations of Quantum Mechanics in the Light of New Technology, pp. 354–368. Physical Society of Japan, Tokyo (1990)
107. Wheeler, J.A.: Recent thinking about the nature of the physical world: it from bit. In: Mendell, R.B., Mincer, A.I. (eds.) Frontiers in Cosmic Physics: Symposium in Memory of Serge Alexander Korff. Annals of the New York Academy of Sciences, vol. 655, pp. 349–364. NY Academy of Sciences, NY (1992)
108. Wheeler, J.A.: At Home in the Universe. American Institute of Physics Publishing, New York (1992)
109. Wheeler, J.A.: Time today. In: Halliwell, J.J., Pérez-Mercader, J., Zurek, W.H. (eds.) Physical Origins of Time Asymmetry, pp. 1–29. Cambridge University Press, Cambridge (1994)
110. Wheeler, J.A., (with Ford, K. W.): Geons, Black Holes, and Quantum Foam: A Life in Physics. W. W. Norton, New York (1998)
111. Wheeler, J.A.: Dated letter to Carroll Alley **13**, (March 1998)
112. Wheeler, J.A.: The eye and the U. Letter dated 19 March 1998 to Abner Shimony, with carbon copies to Arthur Wightman, Peter Mayer, Demetrios Christodoulou, Larry Thomas, Elliot Lieb, Charles Misner, Frank Wilczek, Kip Thorne, Ben Schumacher, William Wootters, Jim Hartle, Edwin Taylor, Ken Ford, Freeman Dyson, Peter Cziffra, and Arkady Plotnitsky
113. Wheeler, J.A.: Information, physics, quantum: the search for links. In: Hey, A.J.G. (ed.) Feynman and Computation: Exploring the Limits of Computers, pp. 309–336. Perseus Books, Reading, MA (1999)
114. Wheeler, J.A.: 'A Practical Tool,' But Puzzling, Too. New York Times (12 December 2000)
115. Zeilinger, A.: The message of the quantum. Nature **438**, 743 (2005)

# Chapter 8
# Toward Physical Realizations
# of Thermodynamic Resource Theories

**Nicole Yunger Halpern**

## 8.1 Introduction

"This is your arch-nemesis."

The thank-you slide of my presentation remained onscreen, and the question-and-answer session had begun. I was presenting a seminar about *thermodynamic resource theories* (TRTs), models developed by quantum-information theorists for small-scale exchanges of heat and work [12, 32]. The audience consisted of condensed-matter physicists who studied graphene and photonic crystals. I was beginning to regret my topic's abstractness.

The question-asker pointed at a listener.

"This is an experimentalist," he continued, "your arch-nemesis. What implications does your theory have for his lab? Does it have any? Why should he care?"

I could have answered better. I apologized that quantum-information theorists, reared on the rarefied air of Dirac bras and kets, had developed TRTs. I recalled the baby steps with which science sometimes migrates from theory to experiment. I could have advocated for bounding, with idealizations, efficiencies achievable in labs. I should have invoked the connections being developed with fluctuation relations [1, 67, 81], statistical mechanical theorems that have withstood experimental tests [4, 6, 9, 10, 42, 51, 53, 66].

The crowd looked unconvinced, but I scored one point: the experimentalist was not my arch-nemesis.

"My new friend," I corrected the questioner.

His question has burned in my mind for two years. Experiments have inspired, but not guided, TRTs. TRTs have yet to drive experiments. Can we strengthen the connection between TRTs and the natural world? If so, what tools must resource theorists develop to predict outcomes of experiments? If not, are resource theorists doing physics?

I will explore answers to these questions. I will introduce TRTs and their role in *one-shot statistical mechanics*, the analysis of work, heat, and entropies on small

N. Yunger Halpern (✉)
Institute for Quantum Information and Matter,
California Institute of Technology, Pasadena, CA 91125, USA
e-mail: nicoleyh@caltech.edu

© Springer International Publishing Switzerland 2017
I.T. Durham and D. Rickles (eds.), *Information and Interaction*,
The Frontiers Collection, DOI 10.1007/978-3-319-43760-6_8

scales. I will discuss whether TRTs can be tested and whether physicists should care. I will identify eleven opportunities for stepping TRTs closer to experiments. Three opportunities concern what we should realize physically and how, in principle, we can realize it. Six adjustments to TRTs could improve TRTs' realism. Two opportunities, less critical to realizations, can diversify the platforms with which we might realize TRTs.

The discussion's broadness evokes a caveat of Arthur Eddington's. In 1927, Eddington presented Gifford Lectures entitled *The Nature of the Physical World*. Being a physicist, he admitted, "I have much to fear from the expert philosophical critic" [25]. Specializing in TRTs, I have much to fear from the expert experimental critic. This paper is intended to point out, and to initiate responses to, the lack of physical realizations of TRTs. Some concerns are practical; some, philosophical. I expect and hope that the discussion will continue.

## 8.2   Technical Introduction

The resource-theory framework is a tool applied to quantum-information problems. Upon introducing the framework, I will focus on its application to statistical mechanics, on TRTs. A combination of TRTs and *one-shot information theory* describes small-scale statistical mechanics. I will introduce select TRT results; more are overviewed in [28].

### 8.2.1   Resource Theories

If you have lived in California during a drought, or fought over the armchair closest to a fireplace, you understand what resources are. *Resources* are quantities that have value. Quantities have value when they are scarce, when limitations restrict the materials we can access and the operations we can perform. *Resource theories* are simple models used to quantify the value ascribed to objects and to tasks by an agent who can perform only certain operations [19]. We approximate the agent as able to perform these operations, called *free operations*, without incurring any cost.

Which quantities have value depends on context. Different resource theories model different contexts, associated with different classes of free operations. Classical agents, for example, have trouble creating entanglement. We can model them as able to perform only local operations and classical communications (LOCC). LOCC define resource theories for entanglement. The resource theory for pure bipartite entanglement is arguably the first, most famous resource theory [35].[1] That theory

---

[1]Conventional thermodynamics, developed during the 1800s, is arguably the first, most famous resource theory. But thermodynamics was cast in explicitly resource-theoretic terms only recently.

has been used to quantify the rate at which an agent can convert copies of a pure quantum state $|\psi\rangle$ into (maximally entangled) Bell pairs. Using a Bell pair and LOCC, one can simulate a quantum channel. Resource states and free operations can simulate operations outside the restricted class.

Resource theorists study transformations between states, such as the quantum states $\rho$ and $\sigma$ or the probability distributions $P$ and $Q$. Can the agent transform $\rho$, resource theorists ask, into $\sigma$ via free operations? If not, how much "resourcefulness" does the transformation cost? How many copies of $\sigma$ can the agent obtain from $\rho^{\otimes n}$? Can $\rho$ transform into a state $\tilde{\sigma}$ that resembles $\sigma$? How efficiently can the agent perform information-processing tasks? What is possible, and what is impossible?

So successfully has the resource theory for pure bipartite entanglement answered these questions, quantum information scientists have cast other problems as resource theories. Examples include resource theories for asymmetry [7, 8, 52], for stabilizer codes in quantum computation [74], for coherence [77], and for randomness [19]. I will focus on resource theories for thermodynamics.

## 8.2.2 Thermodynamic Resource Theories

Given access to a large equilibrated system, an agent ascribes value to out-of-equilibrium systems. Consider a temperature-$T$ heat bath and a hot ($T' \gg T$) gas. By letting the gas discharge heat into the bath, the agent could extract work from equilibration.[2] This work could be stored in a battery, could fuel a car, could power a fan, etc. The resource (the out-of-equilibrium gas) and free operations (equilibration) enable the agent to simulate nonfree operations (to power a car). This thermodynamic story has the skeleton of a resource theory.

The most popular thermodynamic resource theory features an agent given access to a heat bath [12]. The bath has some inverse temperature $\beta := \frac{1}{k_B T}$, wherein $k_B$ denotes Boltzmann's constant. To specify a state $R$, one specifies a density operator $\rho$ and a Hamiltonian $H$ defined on the same Hilbert space: $R = (\rho, H)$. (Hamiltonians are often assumed to have bounded, discrete spectra.) To avoid confusion between $R$ and $\rho$, I will refer to the former as a *state* and to the latter as a *quantum state*.

The free operations, called *thermal operations*, tend to equilibrate states. Each thermal operation consists of three steps: (i) The agent draws from the bath a Gibbs state $G$ relative to $\beta$ and relative to any Hamiltonian $H_b$: $G = (e^{-\beta H_b}/Z, H_b)$, wherein the partition function $Z := \mathrm{Tr}(e^{-\beta H_b})$ normalizes the state. (ii) The agent implements any unitary that commutes with the total Hamiltonian:

$$[U, H_{\mathrm{tot}}] = 0, \quad \text{wherein} \quad H_{\mathrm{tot}} := H + H_b = (H \otimes \mathbb{1}) + (\mathbb{1} \otimes H_b). \quad (8.1)$$

---

[2]More precisely, the agent could extract the capacity to perform work. I omit the extra words for brevity.

This commutation represents energy conservation, or the First Law of Thermodynamics. (iii) The agent discards any subsystem $A$ associated with its own Hamiltonian. Each thermal operation has the form

$$(\rho, H) \mapsto \left( \mathrm{Tr}_A \left( U \left[ \rho \otimes \frac{e^{-\beta H_b}}{Z} \right] U^\dagger \right), H + H_b - H_A \right). \tag{8.2}$$

Each thermal operation decreases or preserves the distance between $R$ and an equilibrium state (for certain definitions of "distance" [32]). As free operations tend to equilibrate states, and as equilibrium states are free, nonequilibrium states are resources. From nonequilibrium states, agents can extract work. Work is defined in TRTs in terms of batteries, or work-storage systems. A battery can be modeled with a two-level *work bit* $B_E = (|E\rangle\langle E|, W|W\rangle\langle W|)$ that occupies an energy eigenstate $|E\rangle \in \{|0\rangle, |W\rangle\}$. By "How much work can be extracted from $R$?" thermodynamic resource theorists mean (in relaxed notation) "What is the greatest value of $W$ for which some thermal operation transforms $R + B_0$ into (Any state) $+ B_W$?" Answers involve one-shot information theory.

### 8.2.3 One-Shot Statistical Mechanics

Statistical mechanics involves heat, work, and entropy. Conventional statistical mechanics describes averages over many trials and over many particles. Yet heat, work, and entropy characterize the few-particle scales that experimentalists can increasingly control (e.g., [15, 18, 27, 46, 59]), as well as single trials. Small scales have been described by TRTs (e.g., [13, 30, 32, 47, 56, 79, 80]). These results fall under the umbrella of *one-shot statistical mechanics*, an application of the *one-shot information theory* that generalizes *Shannon theory* [62].

I will introduce Shannon theory and the *Shannon entropy* $H_S$ with the thermodynamic protocol of compressing a gas. The quantum counterpart of $H_S$ is the *von Neumann entropy* $H_{vN}$. Both entropies, I will explain, quantify efficiencies in the large-scale *asymptotic limit* of information theory, which relates to the *thermodynamic limit* of statistical mechanics. Outside these limits, one-shot entropic quantities quantify efficiencies. I will introduce these quantities, then illustrate them with work performance.

*Shannon theory* concerns averages, over many copies of a random variable or over many trials, of the efficiencies with which information-processing tasks can be performed [21]. Many average efficiencies are functions of the Shannon entropy $H_S$ or the von Neumann entropy $H_{vN}$. Examples include the performance of work.

Consider a classical gas in a cylinder capped by a piston. Suppose that the gas, by exchanging heat through the cylinder with a temperature-$T$ bath, has equilibrated. Imagine compressing the gas quasistatically (infinitely slowly) from a volume $V_i$ to a volume $V_f$. One must perform work against the gas particles that bounce off the piston. The work required per trial averages to the change $\Delta F$ in the gas's *Helmholtz free energy*:

$$\langle W \rangle = \Delta F, \qquad \text{wherein} \qquad F = \langle E \rangle - T S. \tag{8.3}$$

$\langle E \rangle$ denotes the average, over infinitely many trials, of the gas's internal energy; and $S$ denotes the gas's statistical mechanical entropy [75].

The statistical mechanical entropy of a classical state has been cast in terms of the information-theoretic Shannon entropy [38, 75]. The *Shannon entropy* of a discrete probability distribution $P = \{p_i\}$ is defined as

$$H_S(P) := -\sum_i p_i \ln(p_i). \tag{8.4}$$

The Shannon entropy quantifies our ignorance about a random variable $X$ whose possible outcomes $x_i$ are distributed according to $P$. The $P$ relevant to the gas is the distribution over the microstates that the gas might occupy: $P = \{e^{-\beta E_i}/Z\}$, wherein $E_i$ denotes the energy that the gas would have in Microstate $i$ and the partition function $Z$ normalizes the state. The *surprise* $-\ln(p_i)$ quantifies the information we gain upon discovering that the gas occupies Microstate $i$. The surprise quantifies how much the discovery shocks us. Imagine learning which microstate the gas occupies in each of $n$ trials. $H_S(P)$ equals the average, in the limit as $n \to \infty$, of the per-trial surprise. The statistical mechanical entropy $S = k_B H_S(P)$ is proportional to the average of our surprise. $H_S$ quantifies the average, over $n \to \infty$ trials, of the efficiency with which a gas can be compressed quasistatically.

Statistical mechanical averages over $n \to \infty$ trials are related to the *thermodynamic limit*, which is related to the *asymptotic limit* of information theory. The thermodynamic limit is reached as a system's volume and particle number diverge $(V, N \to \infty)$, while $V/N$ remains constant. Imagine performing the following steps $n$ times: (i) Prepare a system characterized by a particular volume $V$ and particle number $N$. (ii) Measure a statistical mechanical variable, such as the energy $E$. Let $E_\gamma$ denote the outcome of the $\gamma$th trial. Let $\langle E \rangle$ denote the average, over $n \to \infty$ trials, of $E$. In the thermodynamic limit, the value assumed by $E$ in each trial equals the average over trials: $E_\gamma \to \langle E \rangle$ [68].

The thermodynamic limit of statistical mechanics relates to an asymptotic limit of information theory. In information theory, the Shannon entropy quantifies averages over $n \to \infty$ copies of a probability distribution $P = \{p_i\}$. A random variable $X$ can have a probability $p_i$ of assuming the value $x_i$. Consider compressing $n$ copies of $X$, jointly, into the fewest possible bits. These $X$'s are called *i.i.d.*, or *independent and identically distributed*: no variable influences any other, and each variable is distributed according to $P$. In the *asymptotic limit* as $n \to \infty$, the number of bits required per copy of $X$ approaches $H_S(P)$. Functions of $H_S$ quantify averages of the efficiencies with which many classical tasks can be performed in the asymptotic limit [21].

The asymptotic average efficiencies of many quantum tasks depend on the *von Neumann entropy* $H_{vN}$ [57]. Let $\rho$ denote a density operator that represents a system's quantum state. The von Neumann entropy

$$H_{vN}(\rho) := -\text{Tr}(\rho \log(\rho)) \tag{8.5}$$

has an operational interpretation like the Shannon entropy's: consider preparing $n$ copies of $\rho$. Consider projectively measuring each copy relative to the eigenbasis of $\rho$. The von Neumann entropy quantifies the average, in the limit as $n \to \infty$, of our surprise about one measurement's outcome [57].

$H_S$ and $H_{vN}$ describe asymptotic limits, but infinitely many probability distributions and quantum states are never processed in practice. Nor are infinitely many trials performed. Into how few bits or qubits can you compress finitely many copies of $P$ or $\rho$? Can you compress into fewer by allowing the compressed message to be decoded inaccurately? How much work must you invest in finitely many gas compressions, e.g., in one "shot"? Such questions have been answered in terms of the *order-$\alpha$ Rényi entropies* $H_\alpha$ and *Rényi divergences* $D_\alpha$ [26, 64].

$H_\alpha$, parameterized by $\alpha \in [0, \infty)$, generalizes the Shannon and von Neumann entropies. In the limit as $\alpha \to 1$, $H_\alpha(P) \to H_S(P)$ (if the argument is a probability distribution), and $H_\alpha(\rho) \to H_{vN}(\rho)$ (if the argument is a quantum state). Apart from $H_1$, two Rényi entropies will dominate our discussion:

$$H_\infty = -\log(p_{\text{max}}) \tag{8.6}$$

depends on the greatest probability $p_{\text{max}}$ in a distribution $P$ or on the greatest eigenvalue $p_{\text{max}}$ of a quantum state $\rho$. $H_\infty$ relates to the work that a TRT agent must invest to create a state $(\rho, H)$ [30, 32]. The work extractable from $(\rho, H)$ relates to

$$H_0 = \log(d) \tag{8.7}$$

wherein $d$ denotes the support of $P$ (the number of nonzero $p_i$'s) or the dimension of the support of $\rho$ (the number of nonzero eigenvalues of $\rho$) [30, 32]. The Rényi divergences $D_\alpha$ generalize the relative entropy $D_1$. They quantify the discrepancy between two probability distributions or two quantum states [26, 57].

Variations on the $H_\alpha$ and the $D_\alpha$ have been defined and applied (e.g., [24, 60, 62, 73]). Examples include the *smooth Rényi entropies* [62] $H_\alpha^\varepsilon$ and the *smooth Rényi divergences* $D_\alpha^\varepsilon$. I will not define them, to avoid technicalities. But I will discuss them in the next section, after motivating smoothing with efficiency.

One-shot entropic quantities quantify the efficiencies with which "single shots" of tasks can be performed. Information-processing examples include data-compression rates [62] and channel capacities [14]. Quantum-information tasks include quantum key distribution [62], the distillation of Bell pairs from pure bipartite entangled states $|\psi\rangle$, and the formation of $|\psi\rangle$ from Bell pairs [11]. Thermodynamic efficiencies include the minimum amount $W_{\text{cost}}$ of work required to create a state $R$ and the most work $W_{\text{yield}}$ extractable from $R$ [32].

The $W_{\text{cost}}$ and $W_{\text{yield}}$ of quasiclassical states have been calculated with TRTs. A *quasiclassical* state $R = (\rho, H)$ has a density operator that commutes with its Hamiltonian[3]: $[\rho, H] = 0$. According to [32], the work extractable from one copy of $R$ is[4] $W_{\text{yield}}(R) = k_{\text{B}}T \, D_0 \left( \rho \, || \, e^{-\beta H}/Z \right)$. Creating one copy of $R$ costs $W_{\text{cost}}(R) = k_{\text{B}}T \, D_\infty \left( \rho \, || \, e^{-\beta H}/Z \right)$.

One-shot efficiencies have been generalized to imperfect protocols whose outputs resemble, but do not equal, the desired states [62]. Suppose that a TRT agent wants to create $R$. The agent might settle for creating any $\tilde{R} = (\tilde{\rho}, H)$ whose density operator $\tilde{\rho}$ is close to $\rho$. That is, the agent wants for $\tilde{\rho}$ to lie within a distance $\varepsilon \in [0, 1]$ of $\rho$: $\mathscr{D}(\tilde{\rho}, \rho) \leq \varepsilon$. (Distance measures $\mathscr{D}$ are discussed in Sect. 8.3.2.4.) $\varepsilon$ is called a *smoothing parameter, error tolerance,* or *failure probability*. The least amount of work required to create any $\tilde{R}$ depends on the smooth order-$\infty$ Rényi divergence $D_\infty^\varepsilon$ [24]: $W_{\text{cost}}^\varepsilon(R) = k_{\text{B}}T \, D_\infty^\varepsilon \left( \rho \, || \, e^{-\beta H}/Z \right)$. The most work extractable from $R$ with a similar faulty protocol depends on the smooth order-0 Rényi divergence: $W_{\text{yield}}^\varepsilon(R) = k_{\text{B}}T \, D_0^\varepsilon \left( \rho \, || \, e^{-\beta H}/Z \right)$ [32].

Let us compare these one-shot work quantities with conventional thermodynamic work. Compressing a gas in the example above costs an amount $W_{\text{cost}}^{\text{th}} = \Delta F$ of work. Consider expanding the gas from $V_f$ to $V_i$. The gas performs an amount $W_{\text{yield}}^{\text{th}} = \Delta F$ of work on the piston. In thermodynamics, the work cost equals the extractable work. In one-shot statistical mechanics, $W_{\text{cost}}(R)$ is proportional to $D_\infty$, whereas $W_{\text{yield}}(R)$ is proportional to $D_0$. From the one-shot results, the thermodynamic results have been recovered [32].

## 8.3 Opportunities in Physically Realizing Thermodynamic Resource Theories

I have collected eleven opportunities from conversations, papers, and contemplation. Opportunity one concerns the question "Which aspects of TRTs merit realization?" Two and three concern how, in principle, these aspects can be tested. I call for expansions of the TRT framework, intended to enhance TRTs' realism, in the next six sections. I invite more-adventurous expansions in the final two sections.

---

[3]Many TRT arguments rely on quasiclassicality as a simplifying assumption. After proving properties of quasiclassical systems, thermodynamic resource theorists attempt generalizations to coherent states.

[4]$D_0$ and $D_\infty$ are called $D_{\text{min}}$ and $D_{\text{max}}$ in [32]. I follow the naming convention in [30]: if $P$ denotes a $d$-element probability distribution and $u$ denotes the uniform distribution $(\frac{1}{d}, \ldots, \frac{1}{d})$, then $D_\infty(P||u) = \log(d) - H_\infty(P)$, and $D_0(P||u) = \log(d) - H_0(P)$.

### 8.3.1 What Merits Realization? How, in Principle, Can We Realize It?

*Prima facie*, single-particle experiments exemplify the one-shot statistical mechanics developed with TRTs. But many-body systems could facilitate tests. I explain how in section one. Section two concerns our inability to realize the optimal efficiencies described by TRT theorems. Section three concerns which steps one should perform in a lab, and how one should process measurements' outcomes, to realize TRT results.

#### 8.3.1.1 What Would Epitomize Realizations of the One-Shot Statistical Mechanics Developed with TRTs?

Many TRT results fall under the umbrella of *one-shot statistical mechanics*, said to describe small scales. In conventional statistical mechanics, small scales involve few particles and small volumes. Yet much of the mathematics of one-shot statistical mechanics can describe many particles and large volumes. To realize TRT results physically, we must clarify what needs realizing. I will first sketch why one-shot statistical mechanics seems to describe single particles. After clarifying in detail, I will demonstrate how one-shot statistical mechanics can describe many-body systems. I will argue that physically realizing one-shot statistical mechanics involves observing discrepancies between Rényi entropic quantities. This argument implies that entangled many-body states could facilitate physical realizations of several one-shot results.

At first glance, "one-shot" appears to mean "single-particle" in TRT contexts. Conventional statistical mechanics describes collections of about $10^{24}$ particles. Most of such particles behave in accordance with averages. (Average behavior justifies our modeling of these collections with statistics, rather than with Newtonian mechanics.) The Shannon and von Neumann entropies ($H_1$) quantify averages over large numbers. Other Rényi entropies ($H_{\alpha \neq 1}$) quantify work in one-shot statistical mechanics. Hence one-shot statistical mechanics might appear to describe single particles.

What one-shot statistical mechanics describes follows from what one-shot information theory describes, since the former is an application of the latter. One-shot information theory governs the simultaneous processing of arbitrary numbers $n$ of copies of a probability distribution $P$ or of a quantum state $\rho$. Hence one-shot statistical mechanics could describe few particles if one copy of $R = (\rho, H)$ described each particle.

One copy can, but need not, describe one particle. Qubits exemplify the "can" statement. The *qubit*, the quantum analog of the bit, is a unit of quantum information. Qubits form the playground in which many one-shot-statistical-mechanics theorems are illustrated (e.g., [48, 81]). A qubit can incarnate as the state of a two-level quantum system. For example, consider an electron in a magnetic field $\mathbf{B} = B\hat{z}$. The Zeeman Hamiltonian $H_Z \propto \sigma_z$ governs the electron's spin, wherein $\sigma_z$ denotes the Pauli $z$-operator. $H_Z$ defines two energy levels, and a density operator $\rho$ represents the electron's spin state. If $R = (\rho, H_Z)$, the one-shot quantities $W_{\text{cost}}(R)$ and $W_{\text{yield}}(R)$ represent the work cost of creating, and the work extractable from, one electron. Hence one copy of a state $R$ can describe one particle; and, as argued above, one-shot statistical mechanics can describe single particles.

Yet one copy of $R$ can represent the state of a system of many particles. For example, let $R = (\rho, H_{XX})$ denote the state of a chain of spin-$\frac{1}{2}$ particles. Such a chain is governed by the Hamiltonian $H_{XX} = -\frac{J_{\text{ex}}}{2} \sum_j (S_j^+ S_{j+1}^- + S_{j+1}^+ S_j^-)$, wherein $J_{\text{ex}}$ denotes the uniform coupling's strength; the index $j$ can run over the arbitrarily many lattice sites; and $S_j^+$ and $S_j^-$ denote the $j^{\text{th}}$ site's raising and lowering operators [53]. This many-body state $R$ can obey TRT theorems derived from one-shot information theory. Hence one-shot statistical mechanics can describe many-body systems.

My point is not that one-shot functions can be evaluated on $n > 1$ i.i.d. copies of a state, denoted by $R^n = (\rho^{\otimes n}, \oplus_{i=1}^n H)$. One-shot functions, such as $H_\alpha$ or $W_{\text{cost}}$, are evaluated on $R^n$ in arguments about the asymptotic limit. Asymptotic limits of one-shot functions lead to expressions reminiscent of conventional statistical mechanics [e.g., Eq. (8.3)], which describes many particles [30, 32, 47]. Therefore, $R^n$ is known to be able to represent the state of a many-particle system. But my point is that $R$, not only $R^n$, can represent the state of a many-particle system. $W_{\text{cost}}(R)$ can equal the work cost of creating "one shot" of a many-particle state [79, 80].

Therefore, few-particle systems would not necessarily epitomize physical realizations of one-shot statistical mechanics. What would? Quantum states $\rho$ whose one-shot entropic quantities ($D_{\alpha \neq 1}, H_0^\varepsilon, H_\infty^\varepsilon$, etc.) were nonzero and differed greatly from asymptotic entropic quantities ($D_1$ and $H_1$) and from each other [63]. I will illustrate with the "second laws" for coherence and with quasiclassical work.

A coherence property of a quantum state $R$, called the *free coherence*, has been quantified with the Rényi divergences $D_\alpha$ [47]. A thermal operation can map $R$ to $S$ only if the $D_\alpha$ decrease monotonically for all $\alpha \geq 0$. In the asymptotic limit, the average free coherence per particle vanishes. Schematically, $D_\alpha/n \to 0$ [47]. Observing coherence restrictions on thermodynamic evolutions—observing a TRT prediction—involves observing nontrivial Rényi divergences.

Second, the approximate work cost $W_{\text{cost}}^\varepsilon(R)$ of a quasiclassical state $R = (\rho, H)$ depends on $D_\infty^\varepsilon(\rho || e^{-\beta H}/Z)$, and the approximate work yield $W_{\text{yield}}^\varepsilon(R)$ depends on $D_0^\varepsilon(\rho || e^{-\beta H}/Z)$. In conventional statistical mechanics, the work cost and the work yield depend on the same entropic quantity. Physically realizing one-shot statistical mechanics involves observing discrepancies between Rényi divergences.

We might increase our chances of observing such a discrepancy by processing a quantum state $\rho$ whose Rényi entropies differ greatly [63]. Example states include

$$\rho_L \to \text{diag}\left(\frac{1}{2}, \underbrace{\frac{1}{2(d-1)}, \ldots, \frac{1}{2(d-1)}}_{d-1}\right), \tag{8.8}$$

which I have represented with a matrix relative to the energy eigenbasis $\{|E\rangle\}$.

Whereas

$$H_\infty(\rho_L) = -\log(p_{max}) = -\log\left(\frac{1}{2}\right) = 1, \qquad (8.9)$$

$$H_0(\rho_L) = \log(d) \gg 1 \qquad (8.10)$$

grows with the dimensionality of the state's Hilbert space. Lacking coherences relative to $\{|E\rangle\}$, $\rho_L$ is quasiclassical. But Stinespring dilation portrays $\rho_L$ as the state of part of a highly entangled many-body system. Not only can one-shot statistical mechanics describe many-body systems, but entangled many-body states might facilitate tests of one-shot theory [63].

Conversely, one-shot theory might offer insights into entangled many-body states. Such states have been called "exotic," and engineering them poses challenges. Suppose that thermal operations modeled the operations performable in a condensed-matter or optics lab. One-shot statistical mechanics would imply which states an experimentalist could transform into which, how much energy must be drawn from a battery to create a state, which coherences could transform into which, etc.

Not all one-shot results could be realized with large systems. A few results depend on the dimensionality of the Hilbert space on which a quantum state is defined. An example appears in [56, App. G.4].[5]

### 8.3.1.2 How Can We Test Predictions of Maximal Efficiencies?

Many TRT theorems concern the maximal efficiency with which any thermal operation can implement some transformation $R \mapsto S$. "Maximal efficiency" can mean "the most work extractable from the transformation," for example. Testing maximal efficiencies poses two challenges: (i) No matter how efficiently one implements a transformation, one cannot know whether a more efficient protocol exists. (ii) Optimal thermodynamic protocols tend to last for infinitely long times and to involve infinitely large baths. These challenges plague not only TRTs, but also conventional thermodynamics. We can approach the challenges in at least three ways: (A) Deprioritize experimental tests. (B) Calculate corrections to the predictions. (C) Check whether decreasing speeds and increasing baths' sizes inches realized efficiencies toward predictions.

---

[5]The authors discuss *catalysis*, the use of an ancilla to facilitate a transformation. Let $R = (\rho, H_R)$ denote a state that cannot transform into $S = (\sigma, H_S)$ by thermal operations: $R \not\mapsto S$. Some *catalyst* $C = (\xi, H_C)$ might satisfy $(\rho \otimes \xi, H_R + H_C) \mapsto (\sigma \otimes \xi, H_S + H_C)$. Catalysts act like engines used to extract work from a pair of heat baths. Engines degrade, so a realistic transformation might yield $\sigma \otimes \tilde{\xi}$, wherein $\tilde{\xi}$ resembles $\xi$. For certain definitions of "resembles," the agent can extract arbitrary amounts of work by negligibly degrading $C$. Brandão et al. quantify this extraction in terms of the dimension $\dim(\mathcal{H}_C)$ of the Hilbert space $\mathcal{H}_C$ on which $\xi$ is defined. The more particles the catalyst contains, the greater the $\dim(\mathcal{H}_C)$. Such one-shot results depend on the number of particles in the system represented by "one copy" of $C$.

Proving that one has implemented a transformation maximally efficiently would amount to proving a negative. Experiments cannot prove such negatives. Imagine an agent who wishes to create some quasiclassical state $R$ from a battery and thermal operations. Suppose that creating any state $\tilde{R}$ that lies within a distance $\varepsilon$ of $R$ would satisfy the agent. The least amount $W_{\text{cost}}^{\varepsilon}(R)$ of work required to create any such $\tilde{R}$ was calculated in [32]. In what fashion could an experimentalist test this prediction, if able to perform arbitrary thermal operations?

The form of a thermal operation that generates a $\tilde{R}$ is implied in [32].[6] One could articulate the operation's form explicitly and could perform the operation in each of many trials. If $\varepsilon$-closeness is defined in terms of the trace distance, one could measure each trial's output, then calculate the distance between the created state and $R$. One could measure the work $W$ invested in each trial and could check whether $W = W_{\text{cost}}^{\varepsilon}(R)$. Separately, one could create $\tilde{R}$'s from another thermal operation, could measure the work $W$ invested in each trial, and could check whether each $W > W_{\text{cost}}^{\varepsilon}(R)$. Finally, one could design a thermal operation $\mathscr{E}$ that outputs a $\tilde{R}$ if sufficient work is invested, could invest $W < W_{\text{cost}}^{\varepsilon}(R)$ in a realization of $\mathscr{E}$, and could check that the resulting state differs from $R$ by more than $\varepsilon$. Yet one would not know whether, by investing $W < W_{\text{cost}}^{\varepsilon}(R)$ in another protocol, one could create a $\tilde{R}$.

Testing optima poses problems also because optimal thermodynamic protocols tend to proceed quasistatically. Quasistatic, or infinitely slow, processes keep a system in equilibrium. Quick processes tend to eject a system from equilibrium, dissipating extra heat [81]. This heat, by the First Law, comes from work drained from the battery [38]. Hence TRT predictions about optimal efficiencies are likely predictions of the efficiencies of quasistatic protocols. Quasistatic protocols last for infinitely long times. Long though graduate students labor in labs, their experiments do not last forever.

Baths' sizes impede experimental tests as time does. In [32, Suppl. Note 1, p. 5], "the energy of the heat bath (and other relevant quantities such as [the] size of degeneracies) [tends] to infinity." Real baths have only finite energies and degeneracies. Granted, baths are assumed to be infinitely large (and unrealistically Markovian) in statistical mechanics outside of TRTs. As such predictions must be reconciled with real baths' finiteness, so must TRT predictions.

The reconciliation can follow one of at least three paths. First, resource theorists can shrug. Shruggers would have responded to the question at my seminar with "No, my arch-nemesis (new friend) might as well not have heard of my results, for all they'll impact his lab." The $W_{\text{cost}}^{\varepsilon}$ prediction appears in the paper "Fundamental limitations for quantum and nanoscale thermodynamics." Fundamental limitations rarely impact experiments more than the imperfectness of experimentalists' control does. For example, quantum noise fundamentally impedes optical amplifiers [17]. Johnson noise, caused by thermal fluctuations, impedes amplifiers in practice, not fundamentally. That is, cooling an amplifier can eliminate Johnson noise. Yet Johnson noise

---

[6]The functional form of $W_{\text{cost}}^{\varepsilon}(R)$ is derived in [32, Suppl. Note 4]. The proof relies on Theorem 2. The proof of Theorem 2 specifies how a TRT agent can perform an arbitrary free unitary. Hence the thermal operation that generates a $\tilde{R}$ is described indirectly.

often outshouts quantum noise. As we cannot always observe fundamental, quantum noise, we should not necessarily expect to observe the fundamental limitations predicted with TRTs.

Nor, one might continue, need we try to observe fundamental limitations. Fundamental limitations bound the efficiencies of physical processes. Ideal bounds on achievable quantities have been considered physics. Thermodynamics counts as physics, though some thermodynamic transformations are quasistatic. TRTs need no experimental tests; ignore the rest of this paper.

Yet thermodynamics has withstood tests [41]. "Quantum-limited" amplifiers uninhibited by Johnson noise have been constructed. Fundamental limitations have impacted experiments, so fundamental limitations derived from TRTs merit testing. Furthermore, testable predictions distinguish physics from philosophy and mathematics. If thermodynamic resource theorists are doing physics, observing TRT predictions would be fitting. Finally, resource theorists motivate TRTs partially with experiments. Experimentalists, according to one argument, can control single molecules and can measure minuscule amounts of energy. Conventional statistical mechanics models such experiments poorly. Understanding these experiments requires tools such as TRTs. Do such arguments not merit testing? If experimentalists observe the extremes predicted with TRTs, then the justifications for, and the timeliness of, TRT research will grow.

Tests of TRT optima can be facilitated by the calculation of corrections and by experimental approaches to ideal conditions. To compute corrections, one could follow Reeb and Wolf [61]. They derive corrections, attributable to baths' finiteness, to Landauer's Principle. [Landauer proposed that erasing one bit of information quasistatically costs an amount $W_L = k_B T \ln(2)$ of work.] The authors' use of quantum-information tools could suit the resource-theory framework. One-shot tools are applied to finite-size baths in [72], and the resource-theory framework is applied in [78].

Instead of correcting idealizations theoretically, one might approach idealizations experimentally. One can perform successive trials increasingly slowly, and one can use larger and larger baths. As speeds drop and baths grow, do efficiencies approach the theoretical bound? Koski et al. posed this question about speed when testing Landauer's Principle. Figure 2 in [44] illustrates how slowing a protocol drops the erasure's cost toward $W_L$. Approaches to the quasistatic limit and to the infinitely-large-bath limit might be used to test TRTs.

### 8.3.1.3   Which Operations Should Be Performed to Test TRT Results?

To test a theorem about a transformation, one should know how to implement the transformation and what to measure. Consider a prediction of the amount of work required to transform a state $R$ into a state $S$. Some thermal operation $\mathscr{E}$ satisfies $\mathscr{E}(R) = S$. Checking the prediction requires knowledge of the operation's form. Which Gibbs state must one create, which unitary must one implement, and which subsystem must one discard? What must be measured in how many trials, and how should the measurement outcomes be processed?

Proofs of TRT theorems detail thermal operations to different extents. Less-detailed proofs include that of Theorem 5 in [80]. The authors specify how a channel transforms a state by specifying the form of the channel's output. Which Gibbs state, unitary, and tracing-out the channel involves are not specified. Farther up the explicitness spectrum lies the proof that, at most, an amount $W_{\text{yield}}^{\varepsilon}(R)$ of work can be extracted with accuracy $1 - \varepsilon$ from a quasiclassical state $R$ [32, Suppl. Note 4]. The proof specifies that "we have to map strings of [weight $1 - \varepsilon$] to a subspace of our energy block." The agent's actions are described, albeit abstractly. The same proof specifies another operation in greater detail, in terms of unitaries: "For each fixed [energy] $E_S$ [of the system of interest,] we apply a random unitary to the heat bath, and identity to [the system of interest]" [32, Suppl. Note 2, p. 7].

One could describe an operation's form more explicitly. For instance, one could specify how strong a field of which type should be imposed on which region of a system for how long. Such details might depend on one's platform—on whether one is manipulating a spin chain, ion traps, quantum dots, etc. Such explicitness would detract from the operation's generality. Generality empowers TRTs. Specifying a Gibbs state, a unitary, and a tracing-out would balance generality with the facilitation of physical realizations.

In particular need of unpacking are the clock, catalyst, coupling, and time-evolution formalisms. Resource theorists developed each formalism, in a series of papers, to model some physical phenomenon. Later papers, borrowing these formalisms, reference the earlier papers implicitly or briefly. The scantiness of these references expedites theoretical progress but can mislead those hoping to test the later results. I will overview, and provide references about, each formalism.

The clock formalism is detailed in [12, 32, 50]. Many TRT theorems concern transformations $(\rho, H) \mapsto (\sigma, H)$ that preserve the Hamiltonian $H$. Theorems are restricted to constant Hamiltonians "without loss of generality." Yet Hamiltonians change in real-life protocols, as when fields $\mathbf{B}(t)$ are quenched. Resource theorists reconcile the changing of real-life Hamiltonians with the frozenness of TRT Hamiltonians by supposing that a clock couples to the system of interest. When the clock occupies the state $|i\rangle$, the Hamiltonian $H(t_i)$ governs the system of interest. The composite system's Hamiltonian remains constant, so TRT theorems describe the composite. TRT theorems restricted to constant $H$'s owe their generality to the clock formalism.

Like the clock formalism, catalysis requires detailing. Laboratory equipment such as clocks facilitates experiments without changing much. These items serve as *catalysts*, according to the resource-theory framework. If TRT predictions are tested in some lab, the catalysts in the lab must be identified. Predictions should be calculated from catalytic thermal operations [13, 56]. If predictions are calculated from thermal operations, the neglect of the catalysts must be justified.

*Prima facie*, couplings do not manifest in TRTs. Free unitaries couple subsystems $R = (\rho, H)$ and $G = (e^{-\beta H_b}/Z, H_b)$. No interaction term $H_{\text{int}}$ appears in $H_{\text{tot}} = H + H_b$. But interaction Hamiltonians couple subsystems in condensed matter and quantum optics. Since condensed matter and quantum optics might provide testbeds for TRTs, the coupling formalisms must be reconciled. An equivalence between the formalisms is discussed in [12, Sect. VIII].

*Prima facie*, TRT systems do not evolve in time. Many authors define states $R = (\rho, H)$ but mention no time evolution of $\rho$ by $U(t) = \exp(-\frac{i}{\hbar} Ht)$. These authors focus on quasiclassical states, whose density operators commute with their Hamiltonians: $[\rho, H] = 0$. This $H$ generates $U(t)$. Hence if $[\rho, U(t)] = 0$, the states remain constant in the absence of interactions, and time evolution can be ignored. Quantum states $\rho$ that have coherences relative to the $H$ eigenbasis evolve nontrivially under their Hamiltonians. This evolution commutes with thermal operations [47, 48]. Though they might appear not to, time evolution and couplings familiar from condensed matter and optics manifest in TRTs.

In addition to the specifying the steps in thermal operations, resource theorists could specify what needs measuring, how precisely, in how many trials, and how to combine the measurements' outcomes, to test TRT predictions. For example, many TRT theorems involve $\varepsilon$-*approximation*. $\varepsilon$-approximation is often defined in terms of the trace distance $\mathscr{D}_{tr}$ between quantum states. $\mathscr{D}_{tr}$ has an operational interpretation. How to use that interpretation, to check the distance from an experimentally created state to the desired state, is discussed below. Instead of invoking the trace distance's operational interpretation, one could perform quantum state tomography. The preciseness with which measurements can be performed must be weighed against the error tolerance $\varepsilon$ in the theorems being tested. As another example, some TRT predictions cannot be tested, as discussed below. We must distinguish which predictions are testable; then specify which measurements, implemented how precisely, and combined in which ways, would support or falsify those predictions.

### 8.3.2   Enhancing TRTs' Realism

Six adjustments could improve how faithfully TRTs model reality. First, we could narrow the gap between the operations easily performable by TRT agents and the operations easily performable by thermodynamic experimentalists. Second, we could incorporate violations of energy conservation into TRTs: TRTs model closed systems, whereas most real systems are open. Section 8.3.2.3 concerns measurements; and Sect. 8.3.2.4 concerns thermal embezzlement, a seeming violation of the First Law of Thermodynamics in TRTs. Faulty operations feature in Sect. 8.3.2.5. Finally, modeling continuous spectra would improve TRT models of classical systems and quantum environments.

#### 8.3.2.1   Should Free Operations Reflect Experimentalists' Capabilities Better? if So, How?

Should resource theorists care about how easily experimentalists can implement thermal operations? Laboratory techniques advance. Even if thermal operations challenge experimentalists now, they might not in three years. Theory, some claim, waits for no experiment. Quantum error correction (QEC) illustrates this perspective. Proposals for correcting a quantum computer's slip-ups involve elaborate architectures, many measurements, and precise control [29]. QEC theory seemed unrealizable when it first flowered. Yet the foundations of a surface code have been implemented [20], and multiple labs aim to surpass these foundations. Perhaps theory should motivate experimentalists to expedite difficult operations.

On the other hand, resource theories are constructed partially to reflect limitations on labs. The agents in resource theories reflect the operationalism inspired partially by experiments. Furthermore, TRTs were constructed to shed light on thermodynamics. Thermodynamics evolved largely to improve steam engines: thermodynamics evolved from practicalities. By ignoring practical limitations, thermodynamic resource theorists forsake one of their goals.[7] Finally, increasing experimental control over nanoscale systems has motivated thermodynamic resource theorists. Experimentalists, an argument goes, are probing limits of physics. They are approaching regimes near the fundamental limitations described by TRTs. This argument merits testing. Physical realizations of TRTs would strengthen the justifications for developing TRTs.

Thermodynamic experimentalists disagree with resource theorists in three ways about which operations are easy: experimentalists cannot easily implement arbitrary unitaries that commute with $H_{tot}$. Nor can experimentalists create arbitrary Hamiltonians. Finally, experimentalists do not necessarily value work as TRT agents do. Redefining thermal operations could remedy these discrepancies.

TRT agents can perform any unitary $U$ that commutes with the total Hamiltonian $H_{tot} = H + H_b$ of a system-and-bath composite [Eq. (8.1)]. Thermodynamic experimentalists cannot necessarily. Consider, for example, condensed matter or quantum optics. Controlling long-range interactions and generating many-body interactions can be difficult. Two-body interactions are combined into many-body interactions. Hence many-body interactions are of high order in two-body coupling constants. Particles resist dancing to the tune of such weak interactions. Coaxing the particles into dancing challenges experimentalists but not TRT agents.

In addition to implementing energy-conserving unitaries, a TRT agent can create equilibrium states ($e^{-\beta H_b}/Z$, $H_b$) relative to the heat bath's inverse temperature $\beta$ and relative to any Hamiltonian $H_b$ (though $H_b$ is often assumed to have a bounded, discrete spectrum). Experimentalists cannot create systems governed by arbitrary $H_b$'s. Doing so would amount to fabricating arbitrary physical systems. Though many-body and metamaterials experimentalists engineer exotic states, they cannot engineer everything.

Most relevantly, experimentalists cannot construct infinitely large baths. Let $\dim(\mathcal{H}_b)$ denote the dimension of the Hilbert space $\mathcal{H}_b$ on which the bath's quantum state is defined. According to the often-used prescription in [32], the output of a thermal channel approaches the desired state $R = (\rho, H)$ in the limit as $\dim(\mathcal{H}_b) \to \infty$.

---

[7]Other goals include the mathematical isolation, quantification, and characterization of single physical quantities. Want to learn how entanglement empowers you to create more states and to perform more operations than accessible with only separable states? Use a resource theory for entanglement. Want to learn how accessing information empowers you? Use the resource theory for information [30, 34].

Real baths' states are not defined on infinitely large Hilbert spaces. Baths' finiteness merit incorporation into TRTs.

Just as experimentalists cannot construct arbitrary bath Hamiltonians $H_b$, experimentalists cannot construct arbitrary ancillary Hamiltonians. Work, for example, is often defined in TRTs with a two-level work bit [32]. Suppose that an agent wishes to transform $R$ into $S$ by investing an amount $W$ of work. The agent borrows a work bit whose gap equals $W$: $B_W = (|W\rangle\langle W|, W|W\rangle\langle W|)$. An experimentalist should prepare $B_W$ to mimic the agent. But an experimentalist cannot necessarily tune a two-level system's gap to $W$. Resource-theory predictions might require recalculating in terms of experimentally realizable ancillas.

Whereas TRT agents can easily implement unitaries and create states that experimentalists cannot, some thermodynamic experimentalists can easily spend work that TRT agents cannot. For example, consider a TRT agent who increases the strength $B(t)$ of a magnetic field $\mathbf{B}(t)$ imposed on a spin system governed by a Hamiltonian $H(t)$. The Hamiltonian's evolution is modeled with the clock formalism in [12, 32]. From this formalism, one can calculate the minimum amount $W_{\text{cost}}$ of work required to increase $B(t)$.

The experimentalist modeled by the agent expends work to increase $B(t)$. One can strengthen a magnetic field by strengthening the current that flows through a wire. One strengthens a current by heightening the voltage drop between the wire's ends, which involves strengthening the electric field at one end of the wire, which involves bringing charges to that end. The charges present repel the new charges, and overcoming the repulsion requires an amount $W'_{\text{cost}}$ of work. Yet thermodynamic experimentalists do not necessarily cringe at the work cost of strengthening a field. They take less pride in charging a battery than in engineering many-body and long-range interactions, in turning a field on and off quickly, and in sculpting a field's spatial profile $\mathbf{B}(\mathbf{r})$. The work cast by resource theorists as valuable has less value, in some thermodynamic labs, than other resources.

Regardless of whether they value work, experimentalists might be able to measure $W'_{\text{cost}}$ [59]. This cost can be compared with the predicted value $W_{\text{cost}}$. TRT results might be tested even if the theory misrepresents some priorities of some thermodynamic experimentalists.[8]

Resource theorists might incorporate experimentalists' priorities and challenges into TRTs. The set of free unitaries, and the set of Hamiltonians $H_b$ relative to which agents can create Gibbs states, might be restricted [40]. One might calculate the cost of implementing a many-body interaction. Corrections might be introduced into existing calculations. Additionally, alternative battery models might replace work bits.

Precedents for altering the definition of free operations exists: thermal operations are expanded to *catalytic thermal operations* in [13, 56]. Experimentalists use engines, clocks, and other equipment that facilitates transformations without altering

---

[8] Such experimentalists value coherence similarly to TRT agents: in experiments, coherent entangled states offer access to never-before-probed physics. In quantum computers, entanglement speeds up calculations. TRT agents value coherence because they can catalyze transformations with coherent states [2]. Agents can also "unlock" work from coherence [43].

(or while altering negligibly). These catalysts, Brandão et al. argue, merit incorporation into free operations. Batteries are similarly incorporated into free operations in [69]. Inspired by [69], Åberg redefined free operations to expose how coherences catalyze transformations [2].

Limitations on experimentalists' control have begun infiltrating TRTs. Brandão et al. showed that each of many thermal operations $\mathcal{E}_1, \mathcal{E}_2, \ldots$ can implement $R \mapsto S$ [12]. If an agent can implement any $\mathcal{E}_i$, the agent can perform the transformation. Performing a particular $\mathcal{E}_i$—controlling particular aspects of the operations—is unnecessary [12, Suppl. Note 7]. Wilming et al. compared TRT agents with agents who can only thermalize states entirely (can only replace states with Gibbs states) [76]. Real experimentalists lie between these extremes: they can perform some, but not all, thermal operations apart from complete thermalization. Following these authors, resource theorists might improve the fidelity with which free operations reflect reality.

### 8.3.2.2 Can Nonconservation of Energy Model Systems' Openness and Model the Impreciseness with Which Unitaries Are Implemented?

TRTs describe closed systems, whose energies are conserved. Every free unitary $U$ commutes with the total Hamiltonian:

$$[U, H_{\text{tot}}] = 0. \tag{8.11}$$

Experimental systems are open and have not-necessarily-conserved energies. We might reconcile the closedness of TRT systems with the openness of experimental systems by modifying the constraint (8.11). A modification could model the impreciseness with which unitaries can be implemented [12].

All subsystems of our universe interact with other subsystems. A quantum system $\mathcal{S}$ in a laboratory might couple to experimental apparatuses, air molecules, etc. Batteries $\mathcal{B}$, clocks $\mathcal{C}$, and catalysts $C$ have been studied in TRTs. The energy of $\mathcal{S} + \mathcal{B} + \mathcal{C} + C$ can more justifiably be approximated as conserved than the energy of $\mathcal{S}$ can. Yet the energy of $\mathcal{S} + \mathcal{B} + \mathcal{C} + C$ can change. If light shines on the clock, photons interact with $\mathcal{C}$. We cannot include in our calculations all the degrees of freedom in a closed, conserved-energy system. To compromise, we might incorporate into TRTs the nonconservation of the energies of $\mathcal{S} + \mathcal{B} + \mathcal{C} + C$.

Brandão et al. introduced nonconservation of energy into TRTs [12, App. VIII]. The authors suppose that work is extracted from $\mathcal{S}$ during $n$ cycles. The noncommutation of $U$ with $H_{\text{tot}}$, they show, corrupts the clock's state. This corruption disturbs the work extraction negligibly in the limit as $n \to \infty$. Even if unable to implement $U$ precisely, the authors conclude, an agent can extract work effectively. The authors' analysis merits detailing and invites extensions to noncyclic processes.

In addition to modeling lack of control, a relaxation of energy conservation would strengthen the relationship between TRTs and conventional statistical mechanics.

As Åberg writes in [Suppl. Note II.A, p. 2] [2], "it ultimately may be desirable to develop a generalization which allows for small perturbations of the perfect energy conservation, i.e., allowing evolution within an energy shell, as it often is done in statistical mechanics."

### 8.3.2.3   How Should TRTs Model Measurements?

The most general process modeled by quantum information theory consists of a preparation procedure, an evolution, and a measurement [57]. A general measurement is represented by a *positive operator-valued measure* (POVM). A POVM is a set $\mathcal{M} = \{M_i\}$ of measurement operators $M_i$. The probability that some measurement of a quantum state $\rho$ yields outcome $i$ equals $\mathrm{Tr}(\rho M_i)$.

Measurements are (postselection is) probabilistic, whereas thermal operations are deterministic. The probability that some thermal operation $\mathcal{E}$ transforms some state $R = (\rho, H_R)$ into some state $S = (\sigma, H_S)$ equals one or zero. Not even if supplemented with a battery can thermal operations implement probabilistic transformations. The absence, from TRTs, of the measurement vertebra in the spine of a realistic protocol impedes the modeling of real physical processes.

The absence also impedes the testing of TRTs. We extract data from physical systems by measuring them. We measure quantum systems through probabilistic transformations. If we attribute to a classical statistical mechanical system a distribution over possible microstates, a measurement of the system's microstate is probabilistic. If TRTs do not model measurements, can we test TRTs?

Alhambra et al. propose a compromise that I will recast [5]. Suppose that an agent wishes to measure a state $\rho' = p\sigma + (1 - p)X$ of a system $\mathcal{S}$, assuming that $\sigma$ and $X$ denote density operators. Suppose that the agent has borrowed a memory $\mathcal{M}$, initialized to a nonfree pure state $(|0\rangle\langle 0|, \mathbb{1}_d)$, from some bank. $\mathcal{M}$ is governed by a totally degenerate Hamiltonian $\mathbb{1}_d$ defined on a $d$-dimensional Hilbert space. A free unitary $U_{\mathrm{record}}$ can couple $\mathcal{S}$ to $\mathcal{M}$. Measuring $\mathcal{M}$ would collapse the state of $\mathcal{S}$ onto $\sigma$ or onto $X$, implementing a quantum POVM.

To incorporate the memory measurement into TRTs, we might add paid-for measurements of quasiclassical states into the set of of allowed operations. The agent should be able to measure a quasiclassical memory, then reset the memory to some fiducial state by investing work.[9] Schematically,

$$(\rho_M, \mathbb{1}_d) + (|W\rangle\langle W|, W|W\rangle\langle W|) \mapsto (|0\rangle\langle 0|, \mathbb{1}_d) + (|0\rangle\langle 0|, W|W\rangle\langle W|). \quad (8.12)$$

The agent originally ascribes to the memory a distribution over possible pure states. Observing which microstate $\mathcal{M}$ occupies, the agent gains information transformable into work [71]. Gaining work deterministically from free operations contradicts the

---

[9]One might worry that $\mathcal{M}$ might not occupy a quasiclassical state after coupling to $\mathcal{S}$. But $\mathcal{M}$ has a totally degenerate Hamiltonian $\mathbb{1}_d$. If $\rho_M$ has coherences relative to the energy eigenbasis, free unitaries can eliminate them [30].

spirit of the Second Law of Thermodynamics. But the agent forfeits this work to erase $\mathcal{M}$. The agent must return $\mathcal{M}$, restored to its original state, to the bank.[10]

Questions about this measurement formalism remain. First, suppose that $U_{\text{record}}$ correlates $\mathcal{S}$ with $\mathcal{M}$ perfectly. The memory's reduced state is maximally mixed, having the spectrum $P_{\mathcal{M}} = (\frac{1}{d} \ldots \frac{1}{d})$. Landauer's Principle suggests that erasing $\mathcal{M}$ costs at least an amount $W = k_{\text{B}}T \log(d)$ of work [45]. Instead, suppose that $U_{\text{record}}$ correlates $\mathcal{S}$ with $\mathcal{M}$ imperfectly. $P_{\mathcal{M}}$ might not be maximally mixed; erasing $P_{\mathcal{M}}$ could cost an amount $W(P_{\mathcal{M}}) < k_{\text{B}}T \log(d)$ of work [30].

Second, measuring and erasing $n$ copies of $\mathcal{M}$ individually costs an amount $n W(P_{\mathcal{M}})$ of work. But the agent might prefer paying a lump sum to paying for each copy individually. In the limit as $n \to \infty$, $(P_{\mathcal{M}})^{\otimes n}$ can be compressed into $n H_{\text{S}}(P_{\mathcal{M}})$ bits. Delaying payment might save the agent work [49]. Delayed payments, like imperfect $U_{\text{record}}$'s, as well as imperfect measurements and nondegenerate memories, merit consideration.

A precedent exists for incorporating paid-for measurements into allowed operations: thermal operations have been expanded to catalytic thermal operations [13, 56]. Just as catalysts are not free, measurements are not: an agent could pay to restore a degraded catalyst to its initial state and can pay to erase a memory. As catalysis has been incorporated into allowed operations, so might measurements. The term $\varepsilon$-deterministic thermal operations has already appeared [67]. $\varepsilon$-determinism surfaces in a related framework, if not in the resource-theory framework, in [1]. Navascués and García-Pintos treat measurement similarly to Alhambra et al. when studying the "resourcefulness" of nonthermal operations [54].

On the other hand, physically realizing TRTs might not necessitate the incorporation of measurements into TRTs. An experimenter could test a work-extraction theorem, derived from the TRT framework, as follows: First, the experimenter extracts work by implementing a thermal operation. Then, the experimenter performs a measurement absent from the TRT framework. The experimenter imitates two agents: One, a TRT agent, extracts work by a thermal operation. This first agent passes the resultant system to someone who "lives outside the resource theory," whom energy constraints do not restrict. The second agent measures the system. Hence one might test TRTs that do not model measurements. But externalizing measurements renders TRTs incomplete models of simple thermodynamic processes. To model physical reality, we must refine the TRT representation of measurements.

---

[10]One might object that the measurement could project the memory's quantum state onto a pure state. Transforming any pure state into $|0\rangle$ costs no work: the transforming unitary commutes with the Hamiltonian $_d$ and so is free [30]. But the agent could be fined the work that one would need to erase $\mathcal{M}$ if one refrained from measuring $\mathcal{M}$. Imagine that a "measurement bank" implements measurements: the agent hands $\mathcal{M}$ to a teller. Depending on the memory's state, the teller and agent agree on a fee, which the agent pays with a charged battery. The teller measures $\mathcal{M}$; announces the outcome; resets $\mathcal{M}$; stores the battery's work contents in a vault; and returns the reset memory $(|0\rangle\langle 0|, _d)$ and the empty battery $(|0\rangle\langle 0|, W|W\rangle\langle W|)$ to the agent.

#### 8.3.2.4 Can a Measure of $\varepsilon$-Closeness Lead to Testable Predictions and Away from Embezzlement?

Many TRT results concern $\varepsilon$-*approximations* (e.g., [5, 13, 30, 32, 54, 56, 79, 80]): suppose that an agent wishes to create a *target state* $R = (\rho, H)$. The agent might settle for some $\tilde{R} = (\tilde{\rho}, H)$ whose density operator $\tilde{\rho}$ lies within a distance $\varepsilon \in [0, 1]$ of $\rho$: $\mathscr{D}(\rho, \tilde{\rho}) \leq \varepsilon$. This $\varepsilon$ has been called the *error tolerance, failure probability*, and *smoothing parameter*. The distance measure $\mathscr{D}$ is often chosen to be the trace distance $\mathscr{D}_{\mathrm{tr}}$. $\mathscr{D}_{\mathrm{tr}}$ has an operational interpretation that could facilitate experimental tests. But $\mathscr{D}_{\mathrm{tr}}$ introduces *embezzlement* into TRTs. Thermal embezzlement is the extraction of arbitrary amounts of work from the negligible degradation of a catalyst. A *catalyst* consists of equipment, such as an engine, that facilitates a transformation while remaining unchanged or almost unchanged. Negligible degradation—the cost of embezzlement—is difficult to detect. Extracting work at a difficult-to-detect cost contradicts the spirit of the First Law of Thermodynamics. Resource theorists have called for eliminating embezzlement from TRTs by redefining thermal operations or by redefining $\varepsilon$-approximation [56, 58, 63]. Three redefinitions have been proposed. On the other hand, embezzlement might merit physical realization.

I will begin with background about the trace distance, defined as follows [57]. Let $\rho$ and $\tilde{\rho}$ denote density operators defined on the same Hilbert space. The *trace distance* between them is

$$\mathscr{D}_{\mathrm{tr}}(\rho, \tilde{\rho}) := \frac{1}{2}\mathrm{Tr}|\rho - \tilde{\rho}|, \qquad (8.13)$$

wherein $|\sigma| := \sqrt{\sigma^\dagger \sigma}$ for an operator $\sigma$. If $\mathscr{D}(\rho, \tilde{\rho}) \leq \varepsilon$, $\rho$ and $\tilde{\rho}$ are called $\varepsilon$-*close*. States $R = (\rho, H)$ and $\tilde{R} = (\tilde{\rho}, H)$ are called $\varepsilon$-*close* if they have the same Hamiltonian and if their density operators are $\varepsilon$-close.

The trace distance has the following operational interpretation [57]. Consider a family of POVMs parameterized by $\gamma$: $\mathscr{M}^\gamma = \{M_i^\gamma\}$. Consider picking a POVM (a value of $\gamma$), preparing many copies of $\rho$, and preparing many copies of some approximation $\tilde{\rho}$. Imagine measuring each copy of $\rho$ with $\mathscr{M}^\gamma$ and measuring each copy of $\tilde{\rho}$ with $\mathscr{M}^\gamma$. A percentage $\mathrm{Tr}(M_i^\gamma \rho)$ of the $\rho$ trials will yield outcome $i$, as will a percentage $\mathrm{Tr}(M_i^\gamma \tilde{\rho})$ of the $\tilde{\rho}$ trials. The difference $|\mathrm{Tr}(M_i^\gamma \rho) - \mathrm{Tr}(M_i^\gamma \tilde{\rho})|$ between the $\rho$ percentage and the $\tilde{\rho}$ percentage will maximize for some $i = i_{\max}$. Imagine identifying the greatest difference $|\mathrm{Tr}(M_{i_{\max}}^\gamma \rho) - \mathrm{Tr}(M_{i_{\max}}^\gamma \tilde{\rho})|$ for each POVM $\mathscr{M}^\gamma$. The largest, across the $\gamma$-values, of these greatest differences equals the trace distance between $\rho$ and $\tilde{\rho}$:

$$\mathscr{D}_{\mathrm{tr}}(\rho, \tilde{\rho}) = \max_\gamma \ |\mathrm{Tr}(M_{i_{\max}}^\gamma \rho) - \mathrm{Tr}(M_{i_{\max}}^\gamma \tilde{\rho})|. \qquad (8.14)$$

This operational interpretation of the trace distance might facilitate tests of TRT results about $\varepsilon$-approximations defined in terms of $\mathscr{D}_{\mathrm{tr}}$.

As an example, consider checking the prediction that, from an amount $W_{cost}^\varepsilon(R)$ of work, thermal operations can generate an approximation $\tilde{R} = (\tilde{\rho}, H)$ to a quasi-classical $R$ [32]. Suppose we identify a thermal operation $\mathscr{E}$ expected to transform the work into a $\tilde{R}$. Suppose we can perform $\mathscr{E}$ in a lab. Suppose we have identified the POVM $\mathscr{M}^\gamma$ that achieves the maximum in Eq. (8.14).[11] We can test the theorem as follows: implement $\mathscr{E}$ on each of many copies of the work resource. Measure, with $\mathscr{M}^\gamma$, the state produced by each $\mathscr{E}$ implementation. Note which percentage of the measurements yield outcome $i$, for each $i$. Prepare many copies of $R$. Measure each copy with $\mathscr{M}^\gamma$. Note which percentage of the measurements yield outcome $i$. Identify the $i$-value $i_{max}$ for which the $R$ percentage differs most from the $\mathscr{E}$ percentage. Confirm that the difference $\Delta$ between these percentages equals, at most, $\varepsilon$.[12]

Though blessed with an operational interpretation, the trace distance introduces embezzlement into TRTs [13, 30, 56]. Suppose that thermal operations cannot transform $R = (\rho, H_R)$ into $S = (\sigma, H_S)$: $R \not\mapsto S$. $R$ might transform into $S$ catalytically. We call $C = (\xi, H_C)$ a *catalyst* if $R \not\mapsto S$ while some thermal operation can transform the composition of $R$ and $C$ into the composition of $S$ and $C$: $(R + C) \mapsto (S + C)$. The catalyst resembles an engine that facilitates, but remains unchanged during, the conversion of heat into work.

Realistic engines degrade:

$$(R + C) \to (S + \tilde{C}). \tag{8.15}$$

Suppose that the Hamiltonians are completely degenerate: $H_R = H_S = H_C = 0$. For every $R$ and $S$, there exist a $C$ and an arbitrarily similar $\tilde{C}$ that satisfy (8.15). The more work the $R$-to-$S$ conversion requires, the larger $C$ must be [the greater the dimension $\dim(\mathscr{H}_C)$ of the smallest Hilbert space $\mathscr{H}_C$ on which $\xi$ can be defined] [13, App. G.2]. The required work is extracted from $C$; this extraction is called *thermal embezzlement*. Embezzlement degrades $C$ to $\tilde{C}$. But if $\dim(\mathscr{H}_C)$ is large enough, the final catalyst state $\tilde{\xi}$ remains within trace distance $\varepsilon$ of $\xi$. Embezzlement also in the context of nondegenerate Hamiltonians has been studied [13, 56].

Embezzlement contradicts the spirit of the First Law of Thermodynamics: embezzlement outputs an arbitrary amount of work at a barely detectable cost. Hence theorists have called for the elimination of embezzlement from TRTs. Three strategies have been proposed: catalysts' sizes and energies might be bounded; $\varepsilon$ might be defined in terms of catalysts' sizes; or $\varepsilon$ might be defined in terms of a distance measure other than $\mathscr{D}_{tr}$. On the other hand, the challenge of realizing embezzlement may

---

[11]Which $\mathscr{M}^\gamma$ achieves the maximum follows from the forms of $R$ and $\tilde{R}$. The experimentalist chooses the form of $R$. The form of the $\tilde{R}$ produced by the thermal operation is described in [32, Suppl. Note 4]. (I have assumed that the $\tilde{R}$ produced by the thermal operation is the state produced in the experiment. But no experiment realizes a theoretical model exactly. This discrepancy should be incorporated into calculations of errors.).

[12]I have ignored limitations on the experimentalist. For example, I have imagined that $\mathscr{E}$ can be implemented infinitely many times and that infinitely many copies of $\rho$ can be prepared. They cannot. These limitations should be incorporated into the error associated with $\Delta$.

appeal to experimentalists, whose violation of the First Law's spirit would appeal to theorists.

Agents cannot borrow arbitrarily large, or arbitrarily energetic, catalysts in [56]. Suppose that an agent wishes to catalyze a transformation in which all Hilbert spaces' dimensions, and all Hamiltonians' eigenvalues, are bounded. The agent in [56] can borrow only catalysts whose Hilbert spaces are small: $\dim(\mathcal{H}_C) \leq d_{\text{bound}}$, for some fixed value $d_{\text{bound}}$. Arbitrary amounts of work cannot be embezzled from such catalysts. Suppose that the agent wants a catalyst $C$ whose spectrum is unbounded (e.g., a harmonic oscillator). Suppose that the partition function $Z_C := \text{Tr}(e^{-\beta H_C})$ associated with the catalyst's Hamiltonian is finite. The catalysts of this sort that the agent can borrow have bounded average energies: $\text{Tr}(\xi H_C) \leq E_{\text{bound}}$. The agent cannot embezzle from these catalysts. In the context of bounded-spectrum catalysts, and in the context of unbounded-spectrum catalysts associated with finite $Z$'s, free operations in [56] are designed to preclude embezzlement.

These limitations on catalysts' sizes and energies have pros and cons. The finite-$Z_C$ assumption "holds for all systems for which the canonical ensemble is well-defined" [56, p. 5]. The assumption seems practical. Yet "there will be specific cases of infinite-dimensional Hamiltonians where simply bounds on average energy do not give explicit bounds on thermal embezzling error" [56, p. 5]. Examples include the hydrogen atom [55]. The restrictions on infinite-dimensional $C$'s do not eradicate embezzling from TRTs. Additionally, bounds on dimension are related to bounds on energy [56], so the two restrictions form a cohesive family. Yet one restriction that eliminated all embezzlement would be more satisfying.

One might eliminate embezzlement from TRTs instead by incorporating catalysts' sizes into the definition of $\varepsilon$-approximation. In [13, App. F4], $\tilde{C}$ is called "$\varepsilon$-close" to $C$ if

$$\mathcal{D}_{\text{tr}}(\xi, \tilde{\xi}) \leq \frac{\varepsilon}{\log(\dim(\mathcal{H}_C))}. \tag{8.16}$$

This definition reflects the unphysicality of arbitrarily large catalysts. Yet the definition destroys the relevance of one-shot information measures to $\varepsilon$-approximate catalysis. Suppose that $\varepsilon$-approximation is defined in terms of the trace distance. Whether $R$ can be $\varepsilon$-catalyzed into $S$ depends on the values of Rényi divergences $D_\alpha$. Now, suppose that $\varepsilon$-catalysis is defined as in Eq. (8.16). Whether $R$ can be $\varepsilon$-catalyzed into $S$ depends only on the relative entropy $D_1$. Testing one-shot statistical mechanics involves the observation of signatures of $D_{\alpha \neq 1}$. Defining $\varepsilon$-approximate catalysis in terms of catalysts' sizes impedes the testing of one-shot theory.

Third, $\varepsilon$-approximation could be defined in terms of distance measures other than the trace distance. Many TRT predictions hold if the definition depends on any contractive metric [30, Sect. 6.1]. The *work distance*, for example, has an operational interpretation and has relevance to one-shot entropies [13, App. G.3]. Suppose that some thermal operation maps $R + C$ to $\widetilde{S + C}$. Suppose that, by investing an amount $\mathcal{D}_{\text{work}}$ of work, one can map $\widetilde{S + C}$ to $S + C$ by a catalytic thermal operation. $\mathcal{D}_{\text{work}}$ is called the *work distance* between $\widetilde{S + C}$ and $S + C$. On the plus side, how much work one can extract from $R + C \mapsto \widetilde{S + C}$ depends on Rényi divergences $D_{\alpha \neq 1}$. On the downside, $\mathscr{D}_{\text{work}}$ lacks information-theoretic properties of distance measures such as $\mathscr{D}_{\text{tr}}$. In terms of which distance measure $\varepsilon$-approximation should be defined remains undetermined.

I have discussed strategies for eliminating embezzlement from TRTs. Embezzlement appears to merit elimination because it contradicts the spirit of the First Law. The spirit, not the letter. Embezzlement seems physically realizable, in principle. Detecting embezzlement could push experimentalists' abilities to distinguish between close-together states $C$ and $\tilde{C}$. I hope that that challenge, and the chance to contradict the First Law's spirit, attracts experimentalists.

#### 8.3.2.5 Can a Definition of $\varepsilon$-Smoothing More Naturally Model Errors and Failure?

The smoothing parameter (alternatively, the error tolerance or failure probability) $\varepsilon$ was introduced in Sect. 8.3.2.4. I detailed the operational interpretation for the $\varepsilon$ defined in terms of the trace distance. This operational interpretation involves an optimal POVM $\mathscr{M}^{\gamma}$. By measuring $\mathscr{M}^{\gamma}$ in many trials, one might verify whether the $S$ outputted by an experimental implementation of some thermal operation is $\varepsilon$-close to the desired state $R$. I mentioned shortcomings of this verification scheme, as well as the embezzlement problem. Another problem plagues $\varepsilon$: Though the "error tolerance" $\varepsilon$ is defined in one-shot information theory, errors seem unable to manifest in single shots of statistical mechanical protocols. Another definition of "error tolerance" might suit TRTs better.

Let us detail how $\varepsilon$ quantifies error in Sect. 8.3.2.4. Suppose you wish to measure, with the optimal $\mathscr{M}^{\gamma}$, each of $n$ copies of $\rho$. Unable to prepare $\rho$ precisely, you create and measure $\tilde{\rho}$ instead. Your measurements' outcomes differ from the ideal measurements' outcomes in a percentage $\varepsilon$ of the trials, in the limit as $n \to \infty$. $\varepsilon$ quantifies the error introduced into your measurement statistics by your substitution of $\tilde{\rho}$ for $\rho$.

The association of $\varepsilon$ with error probability is justified by a limit in which the number $n$ of trials approaches infinity. No error occurs in any single trial; nor does $\varepsilon$ signify the probability that some trial will suffer an error. Yet $\varepsilon$ is defined in one-shot information theory, which describes finite numbers.[13] This paradox suggests that another definition of $\varepsilon$ might suit TRTs better [63].

How can $\varepsilon$ manifest in single shots? Data compression offers an example. Consider compressing a random variable $X$ into the smallest possible number $k$ of bits. Suppose that $X$ has a probability $p_i \neq 0$ of assuming the value $x_i$, for $i = 1, 2, \ldots, d$. One bit occupies one of two possible states. Hence a set of $k$ bits occupies one of $2^k$ possible

---

[13]Granted, one-shot information theory describes the simultaneous processing of finite numbers of *copies of a probability distribution or quantum state*. The finite numbers referred to above are numbers of *sequential trials*. The TRT manifestation of $\varepsilon$ does not contradict one-shot information theory. Yet the former contradicts the spirit of the latter.

states. We associate each possible state of $X$ with one possible state of the bits. Since $X$ assumes one of $d$ possible values, the bits must occupy one of $2^k \geq d$ possible states. We can compress $X$ into, at fewest, $k = \lceil \log(d) \rceil$ bits.

If we used fewer bits, then upon decompressing, we would have some probability of failing to recover the value of $X$. Suppose that $X$ probably does not assume the values $x_1, x_2, \ldots, x_m$. These values' probabilities sum to some tiny number $\varepsilon$: $p_1 + p_2 + \cdots + p_m = \varepsilon$. We can pretend, via a protocol called *smoothing*, that $X$ will assume none of these values [62]. We can compress $X$ almost faithfully into $k' = \lceil \log(d - m) \rceil$ bits. Decompressing, we have a probability $\varepsilon$ of failing to recover the value of $X$, a probability $\varepsilon$ of introducing an error into our representation of $X$. Hence the names *failure probability* and *error tolerance*. Smoothing thus has a natural interpretation in one-shot information-processing tasks.

Smoothing, as I argued, seems to lack a natural interpretation in one-shot statistical mechanics. This lack stems partially from resource theorists' having transplanted the definition of smoothing from information theory into TRTs. Not all transplants bloom in their new environs. Yet smoothing appears relevant to statistical mechanics: Smoothing amounts to ignoring improbable events. Improbable events are ignored in statistical mechanics: Broken eggs are assumed never to reassemble, and smoke spread throughout a room is assumed never to recollect in a fireplace. Since smoothing suits statistical mechanics, but the standard definition of $\varepsilon$ seems unsuited to one-shot statistical mechanics, resource theorists might tailor smoothing to TRTs.

### 8.3.2.6 How Should TRTs Model Continuous Spectra?

Most TRT predictions concern discrete energies. Yet many real physical systems—quantum environments and classical systems—have continuous spectra. We may need to extend TRT predictions to continuous spectra, to model real systems with TRTs.

In many TRT arguments (e.g., [13, 30, 32, 33, 36, 69, 79, 80]), spectra are assumed to be discrete. Simplicity and mathematical convenience justify the assumption. But mathematical convenience trades off with physicality. Resource theorists should check rigorously the limit as the spacing between levels vanishes. TRT predictions have been expected to govern continuous spectra, but limits misbehave.

The need for continuous spectra in TRTs is discussed in [69, 79]. A battery is modeled as a weight that stores free energy as gravitational potential energy. The gravitational-energy operator $H_{grav} = mgx$ depends on the position operator $x$, whose spectrum is continuous. Skrypczyk et al. [70] argue that, if work is extracted from a finite number $n$ of bath qubits, the spectrum of $H_{grav}$ can be approximated as discrete. One can choose for the spacing between consecutive energies to be $\mathscr{E} = \varepsilon/n^2$, for some fixed energy $\varepsilon$. This approximation leads to an error of order $1/n$ [69, App. G]. In the asymptotic limit as $n \to \infty$, the spacing $\mathscr{E}$ vanishes. Whether continuous spectra should be approximated as discrete in other TRT contexts merits investigation.

Incorporating continuous spectra into TRTs could facilitate the modeling of classical systems and quantum environments with which TRTs might be realized physically. Many TRT results concern quasiclassical systems, whose density operators $\rho$ commute with their Hamiltonians $H$: $[\rho, H] = 0$ (e.g., [13, 30, 32, 33, 36, 69, 79, 80]). Such quasiclassical results might be realized more easily than quantum results: Classical platforms, such as DNA and colloidal particles, have been used to test small-scale statistical mechanics such as fluctuation relations [4, 6, 9, 10, 42, 51, 53, 66]. These classical platforms might be recycled to realize TRTs. Classical testbeds may have continuous spectra [39]. Hence we may need to extend quasiclassical theorems to continuous spectra, or to justify the coarse-graining of real systems' continuous spectra into discrete spectra, to harness existing platforms to realize TRTs.

Like classical systems, many environments of open quantum systems have continuous spectra [3]. TRTs model systems coupled to environments—to heat reservoirs [12]. Insofar as realistic reservoirs have continuous spectra, their TRT analogs should. Some reservoirs are modeled with quantum field theories, discussed in Sect. 8.3.3.1, which have continuous spectra. For example, consider a laser mode in a cavity, coupled to the surrounding room by leaky mirrors [3]. This laser is related to the Jaynes-Cummings model, which Åberg introduced into TRTs [2]. The electromagnetic field outside the cavity has a continuum of frequencies. Modeling such QFTs—required to realize TRTs with common systems like lasers—invites us to incorporate continua into TRTs.

## 8.3.3  More-Out-of-the-Way Opportunities

Physical realizations of TRTs require confrontation of the foregoing nine challenges. The following two challenges appear less crucial. Yet the following could lead to realizations with physical platforms that TRTs could not model with just adjustments discussed above. First, modeling quantum field theories with TRTs could facilitate quantum-optics and condensed-matter realizations. Second, TRTs might reach physical platforms via fluctuation relations. Fluctuation relations describe small scales, as TRTs can, and have withstood experimental tests.

### 8.3.3.1  Can TRTs Model Quantum Field Theories?

Quantum field theories (QFTs) represent thermodynamic systems that range from lasers to condensed matter to black holes. Quantum optics and condensed matter are increasingly controllable and conscripted for quantum computation. Holography is shedding the light of quantum information on black holes [31], to which one-shot information theory has been applied [23]. Modeling QFTs with TRTs could unlock testbeds and applications for TRTs. Further motivation appears in [43]: The authors study the "unlocking," aided by an external field, of work stored in coherence. The field suffers a back reaction. Approximating the field as classical neglects the back reaction. To calculate the unlocking's cost, one must quantize the field.

I will overview steps with which we can incorporate QFTs into TRTs: an introduction of Fock space into TRTs, more attention to the number operator in TRTs, and more attention to unbounded spectra. Instead of studying general Hamiltonians and qubits, resource theorists would need to focus on quantum-optics and condensed-matter Hamiltonians. Groundwork for these expansions has been laid in [2, 79, 80].

Specifying a state in a TRT involves specifying a Hilbert space. Quantum states in QFTs are defined on Fock spaces. Since Fock spaces are Hilbert spaces, TRTs offer hope for modeling QFTs. A popular Fock-space basis consists of the eigenstates $|n\rangle$ of the particle-number operator $N$. A similar operator was introduced into TRTs in [79, 80]. The number-like operators in [79, 80] have bounded spectra, whereas arbitrarily many particles can populate a QFT. Spectra were bounded in keeping with the boundedness of the energy spectra in many TRT proofs. TRT spectra are bounded "for convenience" or "for simplicity." This phrasing suggests that proofs are believed to extend easily to unbounded spectra. That the proofs extend merits checking, and how best to model unbounded spectra in TRTs merits consideration. Continuous spectra, their relevance to QFTs, and their incarnation in TRTs is discussed in Sect. 8.3.2.6.

Unbounded spectra and lasers forayed into TRTs in [2, 56, 69]. In [56], catalysts have unbounded spectra. The expectation value of each catalyst's Hamiltonian remains below some finite value $E$. The effective cutoff reduces the problem to a finite-spectrum problem. The cutoff is justified with the finiteness of real systems' energies.

Cutoffs feature also in [69]: a weight (such as a stone) that has gravitational potential energy models a battery. The Hamiltonian $H_{grav} = mgx$ governs the weight, wherein the position operator $x$ denotes the weight's height. The position operator has an unbounded spectrum. Yet the battery's height must lie below some cutoff. Once a weight reaches a certain height above the Earth, $mgx$ approximates the stone's potential energy poorly. Calculating a limit as the cutoff approaches infinity might incorporate truly unbounded spectra fundamentally into TRTs.

In [2], Åberg treats a doubly infinite ladder and a half-infinite ladder as environments. The doubly infinite ladder's energy spectrum runs from $-\infty$ to $\infty$. Though unphysical, the ladder simplifies proofs. Åberg extends these proofs to the more physical half-infinite ladder, whose energy spectrum is bounded from below but not from above. This harmonic oscillator could pave the TRT path toward QFTs, which consist of oscillators that vibrate at various frequencies. One workhorse of QFT is the Jaynes-Cummings model of quantum optics. The model describes how matter exchanges energy with an electromagnetic field. Åberg models a Jaynes-Cummings-like system with his framework. "[T]o what extent such a generalized Jaynes-Cummings interaction can be obtained, or at least approximated, within realistic systems," he writes, "is left as an open question" [2, App. E.2, p. 23].

Åberg discusses also lasers, a stalwart of many labs. Laser light is often modeled with a coherent state,

$$|\alpha\rangle := e^{-\frac{1}{2}|\alpha|^2} \sum_{l=0}^{\infty} \frac{\alpha^l}{\sqrt{l!}} |l\rangle, \qquad (8.17)$$

wherein $\{|l\rangle\}$ denotes the energy eigenbasis and $|\alpha\rangle$ denotes an eigenstate of the annihilation operator $a$: $a |\alpha\rangle = \alpha |\alpha\rangle$ [3]. Åberg studies, rather than coherent states $|\alpha\rangle$, uniform superpositions $\left|\eta_{L,l_0}\right\rangle := \sum_{l=0}^{L-1} \frac{1}{\sqrt{L}} |l_0 + l\rangle$ of neighboring energy eigenstates. Extending his analysis to the $|\alpha\rangle$ can shift TRTs toward modeling lasers and other real systems described by QFTs.

### 8.3.3.2 Can Fluctuation Theorems Bridge TRTs to Experiments?

*Fluctuation theorems* are statistical mechanical predictions about systems arbitrarily far from equilibrium. Fluctuation theorems have withstood experimental tests and describe small scales. Can TRTs reach experiments via fluctuation theorems? If so, how? I will survey progress toward answers.

Fluctuation theorems describe the deviations, from equilibrium values, of outcomes of measurements of statistical mechanical systems. Consider a system coupled to a bath at inverse temperature $\beta$. Suppose that a perturbation changes the system's Hamiltonian from $H_i$ to $H_f$. If the system consists of a gas, it might undergo compression [22]. If the system consist of a trapped ion, a laser might induce a time-evolving field [6]. Consider time-reversing the perturbation. Statistics that characterize the forward process can be related to statistics that characterize the reverse process, to differences $\Delta F$ between free energies, and to deviations from equilibrium statistics. Such relations are fluctuation theorems.

Examples include Crooks' Theorem and Jarzynski's Equality. Let $P_{\text{fwd}}(W)$ denote the probability that some forward trial will require an amount $W$ of work (e.g., to compress the gas). Let $P_{\text{rev}}(-W)$ denote the probability that some reverse trial will output an amount $W$. According to *Crooks' Theorem* [22],

$$\frac{P_{\text{fwd}}(W)}{P_{\text{rev}}(-W)} = e^{\beta(W-\Delta F)}. \qquad (8.18)$$

This $\Delta F = F_f - F_i$ denotes the difference between the free energy $F_f = -k_B T \ln(Z_f)$ of the equilibrium state $e^{-\beta H_f}/Z_f$ relative to the final Hamiltonian and the free energy $F_i$ of the equilibrium state relative to $H_i$. Whereas $\Delta F$ characterizes equilibrium states, $W$ characterizes a nonequilibrium process. Jarzynski's Equality informs us about nonequilibrium, difficult to describe theoretically, via the equilibrium $\Delta F$, easier to describe theoretically. Conversely, from data about nonequilibrium trials, easier to realize in practice, we can infer the value of the equilibrium $\Delta F$, an unrealizable ideal. Multiplying each side of Crooks' Theorem by $P_{\text{rev}}(-W)e^{-\beta W}$, then integrating over $W$, yields *Jarzynski's Equality* [37]:

$$\langle e^{-\beta W} \rangle = e^{-\beta \Delta F}. \qquad (8.19)$$

Fluctuation theorems were derived first from classical mechanics. They have been extended to quantum systems [16] and generalized with information theory (e.g., [65]). Experimental tests have involved DNA [4, 51, 53], trapped colloidal particles [10], single-electron boxes [66], trapped-ion harmonic oscillators [6], and other platforms.

Fluctuation theorems and TRTs describe similar (and, in some cases, the same) problems. First, both frameworks describe arbitrarily-far-from equilibrium processes. Second, each framework features work, entropy, and new derivations of the Second Law. Third, both frameworks describe small scales. I discussed the relationship between small scales and TRTs above. The systems that obey fluctuation theorems most noticeably are small: Deviations from equilibrium behaviors decay with system size [68]. Hence we can most easily detect deviations in small systems, such as single strands of DNA. Fourth, heat exchanges governed by Crooks' Theorem can be modeled with TRT thermal operations [81]. Since the physics described by fluctuation theorems overlaps with the physics described by TRTs, and since experimentalists have tested fluctuation theorems, fluctuation theorems might bridge TRTs to experiments.

Construction of the bridge has begun but needs expansion. Åberg derived a Crooks-type theorem from one-shot statistical mechanics [1, Suppl. Note 10B]. How to model, with TRTs, a process governed by Crooks' Theorem was detailed in [81]. A fluctuation relation was derived from TRT principles in [67]. In [82], one-shot analogs of asymptotic fluctuation-theorem work quantities were derived. These first steps demonstrate the consistency between fluctuation theorems and TRTs (or one-shot statistical mechanics). How to wield this consistency, to link TRTs to experiments via fluctuation theorems, remains undetermined.

## 8.4   Conclusions

During the past few years, the literature about thermodynamic resource theories has exploded. Loads of lemmas and reams of theorems have been proven. To what extent do they describe physical reality? Now that TRTs have matured, they merit experimental probing. I have presented eleven opportunities in physically realizing TRTs. The challenges range from philosophical to practical, from speculative to expectedly straightforward. These opportunities might generalize to physical realizations of other resource theories, such as the resource theory for coherence [77].

I concentrated mostly on gaps in resource theories, on how theorists might nudge their work toward physical realizations. Yet I hope that the discussion will appeal to experimentalists. An experimentalist opened the Q&A of my seminar two years ago. His colleagues and thermodynamic resource theorists have an unprecedented opportunity to inform each other. Let the informing begin... preferably with more cooperation and charity than during the Q&A.

**Acknowledgments**  I am grateful to Fernando Brandão, Lídia del Rio, Ian Durham, Manuel Endres, Tobias Fritz, Alexey Gorshkov, Christopher Jarzynski, David Jennings, Matteo Lostaglio, Evgeny Mozgunov, Varun Narasimhachar, Nelly Ng, John Preskill, Renato Renner, Dean Rickles, Jim Slinkman, Stephanie Wehner, and Mischa Woods for conversations and feedback. This research was supported by an IQIM Fellowship, NSF grant PHY-0803371, and a Virginia Gilloon Fellowship. The Institute for Quantum Information and Matter (IQIM) is an NSF Physics Frontiers Center supported by the Gordon and Betty Moore Foundation. Stephanie Wehner and QuTech offered hospitality at TU Delft during the preparation of this manuscript. I am grateful to Ian Durham and Dean Rickles for soliciting this paper. Finally, I thank that seminar participant for galvanizing this exploration.

# References

1. Åberg, J.: Truly work-like work extraction via a single-shot analysis. Nat. Commun. **4**, 1925 (2013)
2. Åberg, J.: Catalytic coherence. Phys. Rev. Lett. **113**, 15 (2014)
3. Agarwal, G.S.: Quantum Optics. Cambridge U.P. (2013)
4. Alemany, A., Ritort, F.: Fluctuation theorems in small systems: extending thermodynamics to the nanoscale. Europhys. News **41**, 27–30 (2010)
5. Alhambra, Á.M., Oppenheim, J., Perry, C.: What is the probability of a thermodynamical transition?. ArXiv e-prints (2015)
6. An, S., Zhang, J.N., Um, M., Lv, D., Lu, Y., Zhang, J., Yin, Z.Q., Quan, H.T., Kim, K.: Experimental test of the quantum Jarzynski equality with a trapped-ion system. Nat. Phys. **11**, 193–199 (2015)
7. Bartlett, S.D., Rudolph, T., Spekkens, R.W., Turner, P.S.: Degradation of a quantum reference frame. New J. Phys. **8**, 4 (2006)
8. Bartlett, S.D., Rudolph, T., Spekkens, R.W.: Reference frames, superselection rules, and quantum information. Rev. Modern Phys. **79**, 2 (2007). http://dx.doi.org/10.1103/RevModPhys.79.555
9. Bérut, A., Petrosyan, A., Ciliberto, S.: Detailed Jarzynski equality applied to a logically irreversible procedure. Europhys. Lett. **103**(6), 3275–3279 (2013)
10. Blickle, V., Speck, T., Helden, L., Seifert, U., Bechinger, C.: Thermodynamics of a colloidal particle in a time-dependent nonharmonic potential. Phys. Rev. Lett. **96**, 7 (2006). http://link.aps.org/doi/10.1103/PhysRevLett.96.070603
11. Brandao, F.G.S.L., Datta, N.: One-shot rates for entanglement manipulation under non-entangling maps. ArXiv e-prints (2009)
12. Brandão, F.G.S.L., Horodecki, M., Oppenheim, J., Renes, J.M., Spekkens, R.W.: Resource theory of quantum states out of thermal equilibrium. Phys. Rev. Lett. **111**, 25 (2013)
13. Brandão, F.G.S.L., Horodecki, M., Ng, N.H.Y., Oppenheim, J., Wehner, S.: The second laws of quantum thermodynamics. Proc. Nat. Acad. Sci. U.S.A. **112**, 11 (2014)
14. Buscemi, F., Datta, N.: The quantum capacity of channels with arbitrarily correlated noise. ArXiv e-prints (2009)
15. Bustamante, C., Liphardt, J., Ritort, F.: The nonequilibrium thermodynamics of small systems. Phys. Today **58**(7), 43–48 (2005)
16. Campisi, M., Hänggi, P., Talkner, P.: Colloquium: quantum fluctuation relations: foundations and applications. Rev. Modern Phys. **83**(3), 771–791 (2011)
17. Caves, C.M.: Quantum limits on noise in linear amplifiers. Phys. Rev. D **26**(8), 1817–1839 (1982)
18. Cheng, J., Sreelatha, S., Hou, R., Efremov, A., Liu, R., van der Maarel, J.R.C., Wang, Z.: Bipedal nanowalker by pure physical mechanisms. Phys. Rev. Lett. **109**, 23 (2012)

19. Coecke, B., Fritz, T., Spekkens, R.W.: A mathematical theory of resources. ArXiv e-prints (2014)
20. Córcole, A.D., Magesan, E., Srinivasan, S.J., Cross, A.W., Steffen, M., Gambetta, J.M., Chow, J.M.: Demonstration of a quantum error detection code using a square lattice of four superconducting qubits. Nat. Commun. **6** (2015)
21. Cover, T.M., Thomas, J.A.: Elements of Information Theory. Wiley (2012)
22. Crooks, G.E.: Entropy production fluctuation theorem and the nonequilibrium work relation for free energy differences. Phys. Rev. E **60**(3), 2721–2726 (1999)
23. Czech, B., Hayden, P., Lashkari, N., Swingle, B.: The information theoretic interpretation of the length of a curve. ArXiv e-prints (2014)
24. Datta, N.: Min-and max-relative entropies and a new entanglement monotone. IEEE Trans. Inf. Theory **55**(6), 2816–2826 (2009)
25. Eddington, A.S.: The Nature of the Physical World. MacMillan (1929)
26. van Erven, T., Harremoës, P.: R'enyi divergence and kullback-leibler divergence. ArXiv e-prints (2012). http://adsabs.harvard.edu/abs/2012arXiv1206.2459V
27. Faucheux, L.P., Bourdieu, L.S., Kaplan, P.D., Libchaber, A.J.: Optical thermal ratchet. Phys. Rev. Lett. **74**(9), 1504–1507 (1995)
28. Goold, J., Huber, M., Riera, A., del Rio, L., Skrzypczyk, P.: The role of quantum information in thermodynamics—a topical review. ArXiv e-prints (2015)
29. Gottesman, D.: An introduction to quantum error correction and fault-tolerant quantum computation. ArXiv e-prints (2009)
30. Gour, G., Müller, M.P., Narasimhachar, V., Spekkens, R.W., Yunger Halpern, N.: The resource theory of informational nonequilibrium in thermodynamics. Phys. Reports **5380**, 1–58 (2015). http://www.sciencedirect.com/science/article/pii/S037015731500229X
31. Harlow, D.: Jerusalem lectures on black holes and quantum information. ArXiv e-prints (2014)
32. Horodecki, M., Oppenheim, J.: Fundamental limitations for quantum and nanoscale thermodynamics. Nat. Commun. **4**, 1–6 (2013)
33. Horodecki, M., Horodecki, P., Oppenheim, J.: Reversible transformations from pure to mixed states and the unique measure of information. Phys. Rev. A **67**, 6 (2003). http://dx.doi.org/10.1103/PhysRevA.67.062104
34. Horodecki, R., Horodecki, P., Horodecki, M., Horodecki, K., Oppenheim, J., Sen(De), A., Sen, U.: Local information as a resource in distributed quantum systems. Phys. Rev. Lett. **90** (2003)
35. Horodecki, R., Horodecki, P., Horodecki, M., Horodecki, K.: Quantum entanglement. Rev. Modern Phys. **81**, 2 (2009)
36. Janzing, D., Wocjan, P., Zeier, R., Geiss, R., Beth, T.: Thermodynamic cost of reliability and low temperatures: tightening Landauer's principle and the second law. Int. J. Theor. Phys. **39**(12), 2717–2753 (2000)
37. Jarzynski, C.: Nonequilibrium equality for free energy differences. Phys. Rev. Lett. **78**(14), 2690–2693 (1997)
38. Jarzynski, C.: Equilibrium free-energy differences from nonequilibrium measurements: a master-equation approach. Phys. Rev. E **56**(5), 5018–5035 (1997)
39. Jarzynski, C., Quan, H.T., Rahav, S.: The quantum-classical correspondence principle for work distributions. ArXiv e-prints (2015)
40. Jennings, D.: Private communication (2015)
41. Joule, J.P.: On the existence of an equivalent relation between heat and the ordinary forms of mechanical power. Philos. Mag. **xxvii**, 205 (1845)
42. Jun, Y., Gavrilov, M., Bechhoefer, J.: High-precision test of landauer's principle in a feedback trap. Phys. Rev. Lett. **113**, 19 (2014)
43. Korzekwa, K., Lostaglio, M., Oppenheim, J., Jennings, D.: The extraction of work from quantum coherence. ArXiv e-prints (2015)
44. Koski, J.V., Maisi, V.F., Pekola, J.P., Averin, D.V.: Experimental realization of a Szilárd engine with a single electron. Proc. Nat. Acad. Sci. U.S.A. **111** (2014)
45. Landauer, R.: Irreversibility and heat generation in the computing process. IBM J. Res. Dev. **5**, 183–191 (1961)

46. Liphardt, J., Dumont, S., Smith, S.B., Tinoco, I., Bustamante, C.: Equilibrium information from nonequilibrium measurements in an experimental test of Jarzynski's equality. Science (New York, N.Y.) **292**(5574), 1832–1835 (2002)
47. Lostaglio, M., Jennings, D., Rudolph, T.: Description of quantum coherence in thermodynamic processes requires constraints beyond free energy. Nat. Commun. **6** (2015)
48. Lostaglio, M., Korzekwa, K., Jennings, D., Rudolph, T.: Quantum coherence, time-translation symmetry, and thermodynamics. Phys. Rev. X **5**, 2 (2015)
49. Lostaglio, M.: Private communication (2015)
50. Malabarba, A.S.L., Short, A.J., Kammerlander, P.: Clock-driven quantum thermal engines. New J. Phys, **17** (2015)
51. Manosas, M., Mossa, A., Forns, N., Huguet, J., Ritort, F.: Dynamic force spectroscopy of DNA hairpins: II. Irreversibility and dissipation. J. Stat. Mech. Theory Exp. **2** (2009)
52. Marvian, I., Spekkens, R.W.: The theory of manipulations of pure state asymmetry: I. Basic tools, equivalence classes and single copy transformations. New J. Phys. **15**, 3 (2013)
53. Mazza, L., Rossini, D., Fazio, R., Endres, M.: Detecting two-site spin-entanglement in many-body systems with local particle-number fluctuations. New J. Phy. **171**, 013015–013015 (2015). http://adsabs.harvard.edu/abs/2015NJPh...17a3015M. doi:10.1088/1367-2630/17/1/013015
54. Navascués, M., García-Pintos, L.P.: Non-thermal quantum channels as a thermodynamical resource. ArXiv e-prints (2015)
55. Ng, N.: Private communication (2015)
56. Ng, N.H.Y., Mančinska, L., Cirstoiu, C., Eisert, J., Wehner, S.: Limits to catalysis in quantum thermodynamics. ArXiv e-prints (2014)
57. Nielsen, M.A., Chuang, I.L.: Quantum Computation and Quantum Information. Cambridge University Press (2010)
58. Oppenheim, J.: In: Open-Problem Session. Presented at "Beyond i.i.d. in Information Theory 2015". Banff, Canada (2015)
59. Pekola, J.P.: Towards quantum thermodynamics in electronic circuits. Nat. Phys. **11**, 2 (2015)
60. Petz, D.: Quasi-entropies for finite quantum systems. Rep. Math. Phys. **23**(1), 57–65 (1986)
61. Reeb, D., Wolf, M.M.: An improved Landauer principle with finite-size corrections. New J. Phys. **16**, 10 (2014)
62. Renner, R.: Security of Quantum Key Distribution. Ph.D. Thesis (2005)
63. Renner, R.: Private communication (2015)
64. Rényi, A.: On measures of entropy and information. In: Fourth Berkeley Symposium on Mathematical Statistics and Probability, pp. 547–561 (1961)
65. Sagawa, T., Ueda, M.: Generalized jarzynski equality under nonequilibrium feedback control. Phys. Rev. Lett. **104**, 9 (2010)
66. Saira, O.P., Yoon, Y., Tanttu, T., Möttönen, M., Averin, D.V., Pekola, J.P.: Test of the jarzynski and crooks fluctuation relations in an electronic system. Phys. Rev. Lett. **109**, 18 (2012). http://link.aps.org/doi/10.1103/PhysRevLett.109.180601
67. Salek, S., Wiesner, K.: Fluctuations in single-shot $\varepsilon$-deterministic work extraction. ArXiv e-prints (2015)
68. Schroeder, D.V.: An Introduction to Thermal Physics. Addison Wesley, San Francisco, CA (2000)
69. Skrzypczyk, P., Short, A.J., Popescu, S.: Extracting work from quantum systems. arXiv:1302.2811 (2013)
70. Skrzypczyk, P., Short, A.J., Popescu, S.: Extracting work from quantum systems. ArXiv e-prints (2013)
71. Szilard, L.: Über die Entropieverminderung in einem thermodynamischen System bei Eingriffen intelligenter Wesen. Zeitschrift für Physik **53**(11–12), 840–856 (1929)
72. Tajima, H., Hayashi, M.: Refined carnot's theorem. asymptotics of thermodynamics with finite-size heat baths. ArXiv e-prints (2014)
73. Tomamichel, M.: A framework for non-asymptotic quantum information theory. arXiv:1203.2142 (2012)

74. Veitch, V., Hamed Mousavian, S.A., Gottesman, D., Emerson, J.: The resource theory of sta-
    bilizer quantum computation. New J. Phys. **16**, 1 (2014)
75. Vinjanampathy, S., Anders, J.: Quantum thermodynamics. ArXiv e-prints (2015)
76. Wilming, H., Gallego, R., Eisert, J.: Second laws under control restrictions. ArXiv e-prints
    (2014)
77. Winter, A., Yang, D.: Operational resource theory of coherence. ArXiv e-prints (2015)
78. Woods, M.P., Ng, N., Wehner, S.: The maximum efficiency of nano heat engines depends on
    more than temperature. ArXiv e-prints (2015)
79. Yunger Halpern, N.: Beyond heat baths II: framework for generalized thermodynamic resource
    theories. ArXiv e-prints (2014)
80. Yunger Halpern, N., Renes, J.M.: Beyond heat baths: generalized resource theories for small-
    scale thermodynamics. Phys. Rev. E **93**, 022126 (2016)
81. Yunger Halpern, N., Garner, A.J.P., Dahlsten, O.C.O., Vedral, V.: Introducing one-shot work
    into fluctuation relations. New J. Phys. **17** (2015)
82. Yunger Halpern, N., Garner, A.J.P., Dahlsten, O.C.O., Vedral, V.: What's the worst that could
    happen? One-shot dissipated work from R'enyi divergences. ArXiv e-prints (2015)

# Chapter 9
# Merging Contradictory Laws: Imagining a Constructive Derivation of Quantum Theory

William K. Wootters

## 9.1 Introduction

In quantum theory, we compute probabilities by squaring the magnitudes of complex probability amplitudes. Though physicists have gotten used to this procedure, it remains puzzling that nature should be constructed such that this way of computing probabilities agrees so perfectly with the empirical facts. In the decades since the formulation of quantum theory, many authors have aimed to make the theory less mysterious by deriving part or all of its structure from simple axioms (see, for example, [2, 5, 9] as well as several chapters of [6]). However, for me and I think for many others, the essential mystery remains: one feels that there should be a deeper explanation. In this regard it is very much worth noting that the basic structure of quantum theory is independent of the specific physical system being described. For example, the pure states of every binary quantum system can be represented by two-dimensional complex unit vectors, or equivalently, by points on the Bloch sphere. It does not matter whether the system is a photon's polarization, an electron's spin, or two levels of an atom; the description is the same. So if there is a deeper explanation, it must be very general.

The search for a deeper explanation of quantum theory was certainly an abiding interest for John Wheeler, whose work and questions are part of the inspiration behind this volume [18]. Here I explore a particular strategy for deriving the structure of quantum theory not by a direct axiomatic approach but by what I would call a *constructive* approach. The idea is to write down a hypothetical underlying theory from which one hopes quantum theory will emerge mathematically. I hasten to add that in this paper, the effort will not be successful. Nevertheless, I think it is interesting to try to imagine what a constructive derivation might look like. This paper represents one speculative attempt.

W.K. Wootters (✉)
Williams College, Williamstown, MA, USA
e-mail: William.K.Wootters@williams.edu

© Springer International Publishing Switzerland 2017
I.T. Durham and D. Rickles (eds.), *Information and Interaction*,
The Frontiers Collection, DOI 10.1007/978-3-319-43760-6_9

An example of what could be regarded as a constructive derivation of part of quantum theory was given many years ago by Penrose in his work on spin networks [15]. In this work Penrose starts with finite diagrams consisting of lines and vertices, each line being assigned an integer value. Using combinatorial rules, he computes probabilities in a way that, in a certain limit, reproduces the probability rules for the set of pure states of a spin-1/2 particle, which can also be viewed as the set of directions in a three-dimensional space. This is a beautiful piece of work—unlike the work to be presented here, it actually arrives at the correct answer—but as Penrose notes, his combinatorial rules are derived from the irreducible representations of the rotation group in three dimensions, SO(3). He comments that the emergence of directions in a three-dimensional space is surely related to the fact that the rules originated from SO(3). Thus one can still wonder whether it is possible to derive this spin-1/2 structure, and the structure of quantum theory of other systems, from an underlying model that does not in the same way anticipate the theory that is to emerge.

The approach I take here is based on the fact that in order to do physics, we need to make a connection between entities that appear in our physical theory and the objects we experience in the actual world. The core of the paper is a toy model, to be described in detail in Sect. 9.3, in which the identification of theoretical objects with actual objects is the main source of order. The basic idea of the model is as follows. First we decide on a set of admissible deterministic laws, such that each law governs the outcome of any possible experiment. Whenever an experiment is performed—an experiment in our model consists of a preparation followed by a measurement—one of these laws is chosen at random and it determines the outcome of the measurement. If the experiment is repeated, most likely a different law will be chosen. Since the various laws are not consistent with each other, the resulting physics will be probabilistic. Note, though, that in order to apply distinct laws to repetitions of the same experiment, there must be a rule that identifies preparations and measurements as being "the same" across all the laws. In our model, the identification is made in such a way as to maximize the degree of predictability in the resulting physics; that is, the resulting probabilities should be as close as possible to 0 or 1. It is this principle— the maximization of predictability—that gives the model a non-trivial probabilistic structure. In a way, we are invoking here a limited version of Leibniz's assertion that we live in the best of all possible worlds: the world is best in the sense of being the most predictable in the face of an underlying randomness.

In this paper, for the sake of concreteness, I choose as the set of admissible laws what I will call the "classical" laws, that is, laws that are consistent with the notion that measurements simply reveal pre-existing properties. I confess that this choice may not be conducive to deriving quantum theory. It does, however, seem to generate at least a non-trivial probabilistic theory. Conceivably there is a better choice of the set of admissible laws that will actually generate some part of quantum theory.

The following section lays out the basic quantum theory of preparations and measurements. This is the theoretical structure I would like to see emerging from an underlying model. Section 9.3 presents our toy model. The model treats only the case of binary measurements, so it is an attempt to capture some aspect of the quantum mechanics of single qubits. But instead of the usual continuum of possible pure preparations of qubits, in the model there are only finitely many possible preparations and finitely many binary measurements. Ideally, as we increase the number of allowed preparations and measurements, we would see the emergence of the quantum theory of binary systems—that is, we would see the emergence of the Bloch sphere with its associated probabilities. We do not see this. We do seem to see a nontrivial structure emerging, but it is not the Bloch sphere. To a certain degree of approximation, the structure *fits* into the Bloch sphere, and we explore this approximate fitting also in Sect. 9.3. In the final section we discuss the nature of the model and the ways in which its predictions are similar to and different from those of quantum theory.

## 9.2  Quantum Theory of Preparations and Measurements

The basic quantum theory of preparations and measurements for a system with $d$ orthogonal states can be stated as follows:

1. A pure preparation (the only kind we consider in this paper) is represented by a normalized complex vector with $d$ components.
2. A complete orthogonal measurement is represented by an ordered, orthonormal basis for a $d$-dimensional complex vector space. The $d$ vectors in this basis are associated with the $d$ possible outcomes of the measurement.
3. If a system is prepared in the state $|s\rangle$ and then subjected to the measurement $(|m_1\rangle, \ldots, |m_d\rangle)$, the probability of the $j$th outcome is $|\langle s|m_j\rangle|^2$.

Of course there is more to quantum theory than these few rules. But this bare structure is already something one would like to understand more deeply. Now, it is certainly conceivable that the structure we have just described represents nature's lowest level (though it may be expressed in other ways, e.g., in terms of more plausible axioms). There may be no deeper theory. But in this paper we entertain the hypothesis that a lower-level explanation is waiting to be discovered.

For the case of a single qubit—that is, when $d$ equals 2—it is easy to visualize the structure just described. A pure preparation is represented by a point on the unit sphere in three dimensions—the Bloch sphere—and a measurement is represented by a pair of diametrically opposite points on the same sphere. When a preparation represented by the unit vector $\mathbf{a}$ is subjected to a measurement represented by the unit vectors $(\mathbf{b}_1, \mathbf{b}_2)$, the probability of outcome $j$ is $(1/2)(1 + \mathbf{a} \cdot \mathbf{b}_j)$, which is also the squared cosine of half the angle between $\mathbf{a}$ and $\mathbf{b}_j$. In this case, because there are only two outcomes, we can represent a measurement by just the first of the two diametrically opposed unit vectors, the second one then being uniquely determined. So we can express the probability law this way: when a qubit prepared in the state $\mathbf{a}$ is subjected to the measurement $\mathbf{b}$, the probability of the outcome "yes" is $(1/2)(1 + \mathbf{a} \cdot \mathbf{b})$, and the probability of the outcome "no" is $(1/2)(1 - \mathbf{a} \cdot \mathbf{b})$. Surely this is one of the simplest statements one can make within quantum theory. Even seeing this simple statement arising from an underlying model would be highly interesting, as long as the model were compelling and had potential for generalization. Again, we will *not* see such an emergence in this paper. But it is of some interest to see how the model fails.

## 9.3   The Toy Model: Merging Contradictory Classical Laws

We now describe in detail our toy model. Again, the model treats fundamentally binary systems—we want it to capture some aspect of the quantum mechanics of qubits. As we have just seen, in quantum theory there are infinitely many possible preparations of a qubit (one preparation for every point on the Bloch sphere) and infinitely many possible complete orthogonal measurements (again, one for every point on the Bloch sphere). In the toy model, we assume there are exactly $n$ possible preparations and exactly $m$ possible measurements, where $n$ and $m$ are positive integers. An experiment in the model consists of a preparation followed by a measurement, and the outcome of an experiment is either "yes" or "no."

There are two distinct components to our toy model: (i) There is the toy *world*, which consists of $n$ preparation devices and $m$ measuring devices. Each of these devices is labeled, and each persists over time. In what follows, we will use letters near the beginning of the alphabet as labels for the preparation devices, and letters near the end of the alphabet as labels for the measuring devices. (ii) There is the set of possible *classical laws of physics* for this world. A classical law assigns a definite outcome to each possible experiment. Such a law is thus deterministic, but we require more than determinism. We also require that the law be consistent with the following picture: the system being described has exactly two possible states, every preparation prepares the system in one of these two states, and every measurement is a discrimination between these two states. For example, the system might be a coin, with the two possible states heads and tails; every preparation then prepares either heads or tails, and the action of every measurement can be described either as "yes if heads, no if tails" or as "yes if tails, no if heads."

Each such law can be represented by a certain kind of bipartite graph. The graph has two kinds of vertices: those representing preparations and those representing measurements. If a certain preparation-measurement pair would yield the outcome "yes," the corresponding vertices are connected by a line; otherwise they are not. The requirement of classicality means that not every conceivable graph is allowed. Rather, the graph must be such that we can group all the preparation vertices into two categories (the categories could be called "heads" and "tails") and we can group all the measurement vertices into two corresponding categories (which could also be called "heads" and "tails") such that a preparation vertex and a measurement vertex are connected by a line if and only if they are in corresponding categories.

There is a crucial feature of the toy model that I have not yet mentioned: each classical law, like our own laws of physics, does not actually refer to specific objects in the world. That is, the graph that represents a classical law is an *unlabeled* graph (except that the preparation vertices are distinguished from the measurement vertices). Or to

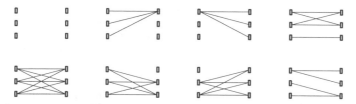

**Fig. 9.1** The eight unlabeled graphs representing classical laws with three possible preparations and three possible measurements. The *blue vertices* on the *left* of each graph represent preparations. The *red vertices* on the *right* represent measurements

put it in other words, what represents a single classical law is an *isomorphism class* of labeled graphs. Figure 9.1 shows all the allowed graphs with three preparation vertices and three measurement vertices (that is, $n = m = 3$).

To get a sense of what these laws mean, consider, for example, the second graph from the left in the upper row. The physical content of this law can be expressed as follows: "There are three preparation devices and three measuring devices. One of the measuring devices yields the outcome 'yes' no matter which preparation device is used. The other two measuring devices yield the outcome 'no' for each preparation device." Note that the law does not tell us which of the actual three measuring devices in the toy world is the special one that yields the outcome 'yes'. Thus, in order to use these laws, we need to assign labels to the vertices. The labels will allow each classical law to be interpreted in terms of the actual objects of the toy world. The assignment of labels is going to be the main source of order in our toy model.

Once the labels are assigned, the toy world operates as follows: when an experiment is performed, one of the classical laws is chosen at random, and it determines the outcome of the experiment.

There are of course many possible ways to put labels on the graphs. One possible labeling for the case $n = m = 3$ is shown in Fig. 9.2. The preparation devices are labeled A, B, C, and the measuring devices are labeled X, Y, Z. In this figure, rather than drawing the graphs exactly as in Fig. 9.1 and putting the labels in different places, we are fixing the locations of the labels and adjusting the lines of the graphs.

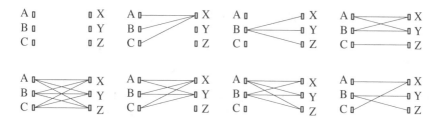

**Fig. 9.2** The same graphs of Fig. 9.1, but now with a particular choice of labels

With the labels assigned, one can now ask questions like this: what is the overall probability that a system prepared with device A and measured with measurement X will yield the outcome "yes"? This probability is computed according to the following prescription:

$$p_{yes}^{AX} = \frac{\text{number of graphs with the outcome "yes" for the experiment AX}}{\text{total number of unlabeled graphs}} \quad (9.1)$$

If the labels are assigned arbitrarily, all the probabilities will tend to be close to 1/2. But if the labeling is done judiciously, at least some of the probabilities can be made closer to 0 or to 1: the world becomes more predictable. With the labeling scheme shown in Fig. 9.2, for example, the probability $p_{AX}$ is equal to 3/4, because in exactly six of the eight graphs, the vertices labeled $A$ and $X$ are connected by a line. The table below shows the probabilities of "yes" for all nine possible experiments in the case $n = m = 3$ if we choose the labeling scheme of Fig. 9.2.

|   | X | Y | Z |
|---|---|---|---|
| A | 3/4 | 1/2 | 1/4 |
| B | 3/4 | 3/4 | 1/2 |
| C | 1/2 | 1/4 | 1/4 |

As I have said in the Introduction, the central idea in our construction is that the degree of predictability should be maximized. We therefore now ask how the labels can be assigned so as to make the world maximally predictable. To make this criterion precise we need to decide on a measure of predictability. We choose to maximize the average "purity" of the binary probability distributions corresponding to the various choices of preparation-measurement pairs—there are $nm$ such pairs in general—where the purity of a binary probability distribution $(p_{yes}, p_{no})$ is defined to be $P = p_{yes}^2 + p_{no}^2$ and $p_{no}$ is simply $1 - p_{yes}$.

For the whole set of 8 graphs of Fig. 9.1, the number of distinct labelings is sufficiently small that we can exhaustively test all the possibilities to find a labeling that maximizes the purity. It turns out that the labeling scheme shown in Fig. 9.2 is optimal. (It is one of many optimal labelings.) It yields an average purity of 7/12, the best possible for $n = m = 3$.

For larger values of $n$ and $m$, it quickly becomes computationally intractable to check every possible labeling, and we resort to a numerical search for a "local" maximum, where "local" means that the average purity cannot be increased by changing the labeling on just one of the graphs. There are typically many local maxima with different values of the average purity; so we try many random initial labelings and report the best of the local maxima that have turned up. Thus the purity value associated with each of the solutions reported in the following subsections should be regarded as a lower bound on the actual optimal purity.

It is worth writing down for all $n$ and $m$ the value of the denominator in Eq. (9.1). This value—that is, the number of unlabeled graphs meeting our classicality condition—is $\lceil (n+1)(m+1)/2 \rceil$: a graph is completely determined by specifying the number of "heads" preparations (this number can go from zero to $n$) and the number of "heads" measurements (this number can go from zero to $m$). So the number of combinations is $(n + 1)(m + 1)$, but we divide by two because we can interchange the roles of "heads" and "tails" without changing the graph. (In general, a graph expresses only *relations* between preparations and measurements.) We take the integer ceiling in case there is a half-and-half graph that is invariant under this interchange.

We consider now two special cases: (i) the case where $n = 3$ (but with different values of $m$), and (ii) the case where $n = m$.

### 9.3.1   The Case $n = 3$

Here we consider the case of just three preparations—we will call them A, B and C as before—and $m$ measurements (with letters near the end of the alphabet as labels). In this case, our numerical calculations suggest that the probabilities emerging from the optimization procedure form a consistent pattern. As an example, we show here a table giving the probabilities we obtain for the case $n = 3, m = 9$. The numbers in the table should all be divided by 20 to get the probabilities (The probabilities in this case can always be expressed as rational numbers with the denominator 20, since there are 20 unlabeled graphs.).

|   | R | S | T | U | V | W | X | Y | Z |
|---|---|---|---|---|---|---|---|---|---|
| A | 5 | 5 | 5 | 5 | 5 | 7 | 9 | 11 | 13 |
| B | 2 | 4 | 6 | 8 | 10 | 12 | 14 | 16 | 18 |
| C | 7 | 9 | 11 | 13 | 15 | 15 | 15 | 15 | 15 |

Thus for preparations A and C, the probabilities have a constant value (1/4 or 3/4) for about half of the measurements, and uniformly spaced values (between 1/4 and 3/4) for the other half. For preparation B the probabilities are uniformly spaced. One sees the analogous pattern for all odd values of $m$ (at least all that I have tried). For even values of $m$ the pattern is almost the same, but there has to be a breaking of symmetry, as illustrated in the following table for $m = 8$ (the numbers in the table are to be divided by 18 to get probabilities).

|   | S | T | U | V | W | X | Y | Z |
|---|---|---|---|---|---|---|---|---|
| A | 5 | 5 | 5 | 5 | 5 | 7 | 9 | 11 |
| B | 2 | 4 | 6 | 8 | 10 | 12 | 14 | 16 |
| C | 6 | 8 | 10 | 12 | 14 | 14 | 14 | 14 |

It happens that tables of this form, whether for odd or even $m$, can always be realized exactly by vectors on the Bloch sphere. We can let the three preparation vectors lie in a plane and be separated by 60°. The measurement vectors are then

**Fig. 9.3** Realizing on the
Bloch sphere the
probabilities obtained by our
optimization procedure for
$n = 3, m = 9$. The *black
arrows* represent preparation
vectors; the *red arrows*
represent measurement
vectors

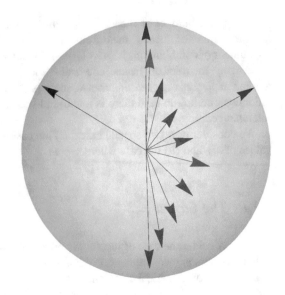

divided into two classes: those at a fixed angle from preparation vector A, and those
at a fixed angle from preparation vector C. Figure 9.3 shows these vectors for the
case $m = 9$.

### 9.3.2 The Case $n = m$

We now let the number of measurements be equal to the number of preparations. In
Table 9.1, we represent a set of probabilities obtained from a numerical search for

**Table 9.1** The numbers in this table, divided by 50, are probabilities obtained by numerically
searching for the maximum average purity for $n = m = 9$

|   | R | S | T | U | V | W | X | Y | Z |
|---|---|---|---|---|---|---|---|---|---|
| A | 30 | 35 | 40 | 41 | 40 | 37 | 34 | 31 | 28 |
| B | 20 | 25 | 30 | 35 | 38 | 39 | 38 | 37 | 36 |
| C | 12 | 17 | 22 | 27 | 32 | 37 | 40 | 41 | 42 |
| D | 8 | 13 | 18 | 23 | 28 | 33 | 38 | 41 | 44 |
| E | 5 | 10 | 15 | 20 | 25 | 30 | 35 | 40 | 45 |
| F | 6 | 9 | 12 | 17 | 22 | 27 | 32 | 37 | 42 |
| G | 8 | 9 | 10 | 13 | 18 | 23 | 28 | 33 | 38 |
| H | 14 | 13 | 12 | 11 | 12 | 15 | 20 | 25 | 30 |
| I | 22 | 19 | 16 | 13 | 10 | 9 | 10 | 15 | 20 |

Each row corresponds to a preparation device, each column to a measuring device the case $n = m = 9$. The probabilities are obtained from the values in the table by dividing by 50.

The pattern is perhaps not as obvious as in the preceding subsection, but there certainly is a discernible structure in the table. For example, along the rows one sees many sequences with uniform increments. In contrast with the case $n = 3$, here it seems not to be the case that the probabilities can be realized by vectors in the Bloch sphere. However, such vectors can *approximate* these probabilities reasonably well. Figure 9.4 shows vectors on the Bloch sphere for which the probabilities differ from the ones of Table 9.1 with a root-mean-square deviation of 0.016. In making this picture I forced the measurement vectors to lie in a plane, simply because it turns out that even without this restriction the vectors tend to be very nearly in the same plane. (Without this restriction one can actually match the probabilities of the table somewhat better, with a root-mean-square deviation of 0.010, though the picture does not look much different.) In contrast, for a $9 \times 9$ table of *randomly chosen* probability values (each chosen from the uniform distribution over the interval $[0, 1]$), the closest approximation one can obtain with Bloch vectors typically deviates from the actual probabilities with a root-mean-square deviation larger than 0.15.

Note that even though, when $n$ equals $m$, the basic mathematical optimization problem is completely symmetric between preparations and measurements, the solution apparently has to break the symmetry. In Table 9.1, there is a row exhibiting strictly uniform increments (the row labeled E) but no column with this property. In other optimization trials, the roles of rows and columns are interchanged.

One might wonder whether the optimal probabilities for $n = m$ come closer to being realizable by means of Bloch vectors as the value of $n$ increases. It is hard to know, but the evidence does not suggest that this is the case. Figure 9.5 shows a plot of the root-mean-square deviation between the probabilities obtained from the toy

**Fig. 9.4** Bloch vectors that approximately realize the probabilities we obtain for the case $n = m = 9$. The *black arrows* represent preparation vectors, in the order A through I from *top* to *bottom*. The *red arrows* represent measurement vectors, in the order R through Z from *left* to *right*

**Fig. 9.5** A plot of the root-mean-square deviation between the probabilities obtained from the toy model, for $n = m$, and the closest probabilities we have found that can be obtained from vectors on the Bloch sphere

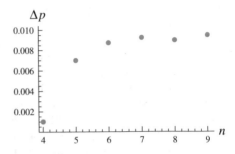

model and the closest Bloch-vector probabilities we have found. The plot does not suggest that this deviation will approach zero for large $n$.

Of course, even if one could realize the probabilities of the toy model with Bloch vectors, one could not claim that the toy model reproduces the quantum theory of qubits. For that, we would want both the preparation vectors and the measurement vectors to be roughly uniformly distributed over the Bloch sphere, approaching perfect uniformity as $n$ approaches infinity. The toy model does not do anything like this. Rather, to the extent that the probabilities can be approximated by Bloch-vector probabilities, the vectors themselves, for both the preparations and the measurements, appear to lie on one-dimensional curves rather than being spread roughly uniformly over the surface of the sphere. Indeed, it could well be that with the set of *classical* laws, one will not be able to generate a nearly uniform distribution of state vectors and measurement vectors over the Bloch sphere, no matter how the vertices are labeled.

We have not at all discussed here the *time evolution* of quantum states. If it had turned out that the measurements and preparations could be mapped naturally and uniformly onto the Bloch sphere, then one can imagine ways in which an extension of the model could have led to the concept of evolution by unitary transformations, since unitary transformations preserve angles on the Bloch sphere. But the structure that does emerge from our toy model does not have this kind of symmetry. The measurement vectors shown in Fig. 9.4 lie in a plane, but they lie in only half of the plane, and they are not uniformly spaced. One aspect of the lack of symmetry is that certain preparations, and certain measurements, are more likely than others to produce the "yes" outcome overall. For example, in Table 9.1, preparation A is the one most likely to produce a "yes" outcome, averaged over all measurements, and measurement Z has the analogous property among measurements. This lopsidedness is unlike the situation in quantum theory, in which there is no preferred vector, and no preferred basis, in Hilbert space.

There is a natural extension of our toy model to systems with more than two perfectly distinguishable states. Let $d$ be the number of such states. (If $d = 6$, we can think of the toy system as an ordinary die, but the goal would be to model a quantum system with Hilbert-space dimension 6.) We take the preparations to be pure preparations and the measurements to be complete orthogonal measurements. A deterministic law can be expressed as a generalized bipartite graph that has preparation vertices and measurement vertices as before, but with each preparation now related to each measurement in one of $d$ ways (rather than in one of two ways). The condition of classicality can be expressed in this way: the preparation vertices can be partitioned into $d$ equivalence classes (the null class being an allowed possibility), and the measurement vertices can be partitioned into $d!$ equivalence classes, such that the relation between any two measurement classes is characterized by a permutation of the preparation classes. (In the case of an ordinary die, one can think of a measurement as assigning an outcome label to each of the die values $1, 2, \ldots, 6$. But different measurements can make different correspondences between the die values and the outcome labels.) Each unlabeled classical graph would then be given a specific labeling, with the labels chosen so as to maximize the average purity over all the possible experiments. As before, whenever an experiment is performed, one of these graphs would be chosen at random and would determine the outcome of the experiment. I have not yet carried out any numerical trials to see how the optimization works out for this more general problem.

## 9.4 Discussion

We have explored a simple toy model that generates the probabilities of the outcomes of hypothetical binary experiments, where each experiment consists of a preparation followed by a measurement. The model appears to yield a non-trivial structure, as exemplified in the $9 \times 9$ table of the preceding section. This structure shares certain qualitative features with the quantum theory of qubits: (i) the probabilities lie between the completely deterministic case, in which all probabilities are either 0 or 1, and the completely random case, in which all probabilities are equal to 1/2; (ii) if two preparations both yield probabilities close to 1 for any one measurement (or both close to 0), then they tend to yield very similar probabilities for all other measurements. (See rows D and E of Table 9.1, for example.) However, in its details the structure produced by the toy model does not particularly resemble the quantum theory of qubits. It does not yield the Bloch sphere.

Nevertheless, it is interesting to think about what sort of model we are dealing with. Spekkens and Harrigan have distinguished two kinds of hidden-variable model that one might consider in an attempt to look for a theory underlying quantum theory: $\psi$-ontic models and $\psi$-epistemic models [11]. In a $\psi$-ontic model, the quantum state is taken to be an element of reality, but in addition there are other variables that have some bearing on the results of measurements. The Bohm model is of this type [3, 4]. In a $\psi$-epistemic model, on the other hand, the quantum state is not an element of reality but instead represents a probability distribution over more fundamental properties. In such a model, quantum probabilities are epistemic probabilities: we do not know the actual, ontic state of the system being measured, and the probabilities simply reflect our ignorance [17].

In the toy model explored in this paper, there is something hidden: the set of labeled graphs is not directly observable but is manifested only in the probabilities of the outcomes of the various experiments. Moreover, observers in the toy world do not know which graph will be in effect for any given experiment. But the model is not a hidden-variable model in the usual sense and does not fit into either of the Spekkens-Harrigan categories. In our model the analog of a quantum state would be a single row of the probability table (such as Table 9.1). Such a row of probabilities is not a fundamental object in the theory—it is secondary and derived. But neither is it the case that each row—each "quantum state"—represents a different probability distribution over the same set of ontic states, as would be the case in a $\psi$-epistemic model. Rather, all the rows are partial expressions of the *same* underlying probability distribution, namely, the uniform distribution over a certain set of labeled graphs, e.g., the graphs of Fig. 9.2.

Note that this underlying probability distribution entails certain *correlations* among the outcomes of different experiments. For example, the uniform distribution over the graphs of Fig. 9.2 entails a correlation between the experiment CZ and the experiment AX: the experiment CZ never yields the outcome "yes" in a graph for which AX yields the outcome "no." Whether or not this kind of correlation should be regarded as observable depends on how one fleshes out the toy model. If, as we have been suggesting, a new random graph is chosen every time a new experiment is performed, then the correlations have no operational meaning: they cannot be observed. On the other hand, one can imagine a different toy world in which a particular choice of classical law affects more than one experiment. For example, in a world in which time is discrete and universal, it could be the case that all experiments completed at the same moment are governed by the same classical law. Then, in addition to the basic probability law, there would be strange correlations among experiments simply because they were completed at the same time.

A number of theorems have been proven showing that certain classes of $\psi$-epistemic theories are incompatible with the predictions of quantum theory [1, 7, 8, 10, 13, 14, 16]. (For an extensive review see Ref. [12].) But at least as regards modeling the quantum mechanics of single systems, there can be no general no-go theorem for a model based on a probability distribution over deterministic laws, as I explain in the following paragraph. (There *could* be a no-go theorem if the laws are required to be classical in our sense. Moreover, the situation is more subtle for multipartite quantum systems. We would need to give specific rules in the toy world that apply to a measurement made on part of a multicomponent system.)

For simplicity let us consider first the qubit case, which has been our main concern in this paper. It is certainly possible to come up with a set of labeled bipartite graphs, like those of Fig. 9.2 but with many more vertices, that approximates arbitrarily well the actual qubit probabilities. First choose a set of points on the Bloch sphere that cover the surface fairly uniformly. These points can represent the preparations, and they can also represent the measurements. Now construct a series of bipartite labeled graphs, each one generated by randomly assigning "yes" or "no" to each preparation-measurement pair in accordance with the quantum probabilities. (These graphs will typically not be "classical.") With enough graphs generated in this way, the frequencies of occurrence of "yes" and "no" for each preparation-measurement pair will typically be very close to the actual quantum probability, simply because the quantum probabilities were used to generate the graphs. It is clear that an analogous construction can be done for higher-dimensional quantum systems as well.

Of course one could not regard this way of reproducing the quantum probabilities as any kind of *explanation* of the quantum probabilities. One would simply be assuming what one was trying to explain. I offer this method only as a way of showing that a collection of deterministic laws *can* reproduce the quantum probabilities. The trick would be to do this in a way that does not assume the answer. Again, the motivation for this work is not primarily to interpret quantum probabilities. Rather, it is to try to imagine how the familiar quantum theoretical structure—that pure states are complex vectors, and that probabilities are squared magnitudes of inner products—arises.

It is an interesting exercise to imagine ourselves living in a world where the probabilities of the outcomes of experiments were actually given by the numbers that emerge from our toy model in the limit of a large number of preparations and measurements. (Since these probabilities do not seem to agree with the quantum probabilities, we would not be living in our familiar quantum world.) With enough thought, we who are living in this hypothetical world might come up with a simple mathematical characterization of the probability laws. For binary measurements we would not get the Bloch sphere but some other structure. However, knowing this structure would not necessarily reveal to us that the probabilities were actually the result of an optimization over all possible labelings of a set of graphs. In the same way, it is conceivable that in our own world, though there is a certain simplicity in the quantum rules as they stand, those rules themselves may actually emerge from something even simpler that we are failing to see.

Of course even if such a deeper layer exists, it will surely be very hard simply to guess the underlying structure. In a way this paper presents one guess, which turns out, not surprisingly, to be wrong. Almost all guesses will be wrong. If there is a deeper level underlying quantum theory, we may not be able to find it until we begin to see experimental deviations from quantum theory that give us some clue as to where we should look.

Finally, it is worth recalling the words with which John Wheeler anticipated the discovery of a deeper understanding of quantum theory: "Behind it all is surely an idea so simple, so beautiful, so compelling that when—in a decade, a century, or a millennium—we grasp it, we will all say to each other, how could it have been otherwise? How could we have been so stupid for so long?" [18].

**Acknowledgements** I am grateful to Chris Fuchs for comments on an earlier draft. This research was supported by the Foundational Questions Institute (grant FQXi-RFP3-1350).

# References

1. Aaronson, S., Bouland, A., Chua, L., Lowther, G.: Psi-epistemic theories: the role of symmetry. Phys. Rev. A **88**, 032111 (2013)
2. Birkhoff, G., von Neumann, J.: The logic of quantum mechanics. Ann. Math. **37**, 823–843 (1936)
3. Bohm, D.: A suggested interpretation of the quantum theory in terms of "hidden" variables. I. Phys. Rev. **85**, 166–179 (1952)
4. Bohm, D.: A suggested interpretation of the quantum theory in terms of "hidden" variables. II. Phys. Rev. **85**, 180–193 (1952)
5. Chiribella, G., D'Ariano, G.M., Perinotti, P.: Informational derivation of quantum theory. Phys. Rev. A **84**, 012311 (2011)
6. Chiribella, G., Spekkens, R.W. (eds.): Quantum Theory: Informational Foundations and Foils. Springer, Dordrecht (2016)
7. Colbeck, R., Renner, R.: Is a system's wave function in one-to-one correspondence with its elements of reality? Phys. Rev. Lett. **108**, 150402 (2012)
8. Colbeck, R., Renner, R.: A system's wavefunction is uniquely determined by its underlying physical state. arXiv:1312.7353 (2013)
9. Hardy, L.: Quantum Theory From Five Reasonable Axioms. arXiv: quant-ph/0101012 (2001)
10. Hardy, L.: Are quantum states real? Int. J. Modern Phys. B **27**, 1345012 (2013)
11. Harrigan, N., Spekkens, R.W.: Einstein, incompleteness, and the epistemic view of quantum states. Found. Phys. **40**, 125 (2010)
12. Leifer, M.: Is the quantum state real? An extended review of $\psi$-ontology theorems. Quanta **3**, 67 (2014)
13. Mansfield, S.: Reality of the quantum state: a stronger psi-ontology theorem. arXiv:1412.0669 (2014)
14. Montina, A.: Communication complexity and the reality of the wave-function. Modern Phys. Lett. A **30**, 1530001 (2015)
15. Penrose, R.: Angular momentum: an approach to combinatorial space-time. In: Bastin, T. (ed.) Quantum Theory and Beyond, pp. 151–180. Cambridge Univercity Press, London (1971)
16. Pusey, M.F., Barrett, J., Rudolph, T.: On the reality of the quantum state. Nat. Phys. **8**, 475 (2012)
17. Spekkens, R.W.: In defense of the epistemic view of quantum states: a toy theory. Phys. Rev. A **75**, 032110 (2007)
18. Wheeler, J.A.: How Come the Quantum? New Techniques and Ideas in Quantum Measurement Theory. Ann. N. Y. Acad. Sci. **480**, 304–316 (1986)

# Chapter 10
# Understanding the Electron

Kevin H. Knuth

## 10.1 Introduction

Whether it is the crack and snap of an electric shock on a cold winter day or the boom and crash of a lightning bolt on a stormy summer afternoon, we are familiar with electrons because they influence us. Similarly, scientists know about electrons because they influence their measurement equipment. Electrons are described in terms of properties inspired by our descriptions of billiard balls, such as mass, position, energy, and momentum, as well as additional non-billiard-ball-like properties such as charge and spin. The laws of physics, as applied to electrons, focus on describing the interrelationships among these relevant variables as well as their relationships to external forces.

Since the discovery of the electron [52], we have become so familiar with these electron properties that we have come to view them as foundational. This is despite the fact that there are well-known serious conceptual problems that have plagued even the most prominent physicists over the last century. For example, Einstein, when faced with the burgeoning field of particle physics, was compelled to write, "You know, it would be sufficient to really understand the electron." This is a difficult issue since quantum mechanics tells us that in some situations an electron acts as a particle and in others it acts as a wave. Moreover, quantum mechanics also tells us that an electron cannot simultaneously possess both a definite position and a definite momentum. While this may not be so surprising given the familiarity that today's physicists have with quantum mechanics and the fact that the vast majority have simply come to accept that this is how things are, the conceptual issues remain unsolved. Now it is perhaps less well known that Breit [8] and Schrödinger [47] both independently showed that the Dirac equation, which most accurately describes the behavior of a single electron, results in velocity eigenvalues of $v = \pm c$. That is, at

K.H. Knuth (✉)
University at Albany, Albany, NY, USA
e-mail: kevin.h.knuth@gmail.com

© Springer International Publishing Switzerland 2017
I.T. Durham and D. Rickles (eds.), *Information and Interaction*,
The Frontiers Collection, DOI 10.1007/978-3-319-43760-6_10

the finest of scales, the electron can *only* be observed to move the speed of light! Schrödinger called this phenomenon *Zitterbewegung*, or shuddering motion, since the implication is that an electron must either zig-zag back-and-forth or spiral at the speed of light—even when at rest [23–27]. Not only does this bring into question what is meant by the state of rest, but it is also seemingly contrary to what we expect from a massive particle within the context of special relativity. And yet, the Dirac equation describes the relativistic quantum mechanics of a single electron. However, we do not need to go so far as quantum mechanics to identify lingering conceptual problems. For instance, most of the electron properties are observer-dependent: position, time, speed, energy, momentum, component of spin, which strongly suggests that they are not *properties possessed* by an electron, but rather they are somehow a *description of the relationship* between the electron and the observer.

Physicists studying foundations must carefully consider the implications of such clues and conceive of theoretical models that accommodate them. This is not accomplished by assuming high-level concepts such as Lorentz invariance, Hilbert spaces, and Lagrangians, since there exists no foundational rationale for such detailed technical assumptions—save for the current conceptual formulation of physics, which we aim to understand more deeply. Likewise, principles involving the relationships among relevant variables, such as energy and momentum in the Principle of Conservation of Four-Momentum or position and momentum in the Principle of Complementarity, will not serve us either since the adoption of such principles precludes us from going deeper and understanding the nature of the relevant variables themselves. One cannot understand why something is true by assuming it to be true. Instead, we must ask the question: What is the nature of the electron properties such that they behave the way they do, and is there perhaps a simpler way to think about them?

In this paper, I present a simple model of the electron and show that it faithfully reproduces a surprising amount of physics. It would be naive to assume that the ideas presented here comprise the final word on the matter. Instead, it is hoped that this work will demonstrate that a simple and understandable picture of elementary particles, as well as a foundation for physics, is indeed feasible and should be actively sought after.

## 10.2 Electron Properties

Let us begin by performing a thought experiment. Imagine that electrons are pink and fuzzy, but that their pinkness and fuzziness do not affect how they influence any measurement apparatus. Since these hypothetical electron properties are undetectable by measurement, they are forever inaccessible to us as experimenters. That is, there is no way for us to know about the pink and fuzzy nature of the electron. Turning this thought on its side, it becomes readily apparent that the only properties of an electron that we as experimenters have access to are those that affect how an electron influences our equipment. In fact, an operationalist definition of such properties would consist simply of a description of the effects of the electron's influence.

This is, in fact, how one builds up the development of the theory of magnetism in a class lecture. The magnetic force is defined by how charges act in a variety of situations. We now know, since relativity, that the magnetic force itself is observer-dependent in that in a different reference frame it can be perceived as a combination of an electric and magnetic force, or in the limiting case, simply an electric force. For this reason, the modern perspective involves the concept of the electromagnetic force where the electric or magnetic nature of the force is dependent on the relationship between the system and the observer.

The observation that many electron properties are observer-dependent suggests that what we think of as properties of the electron more accurately reflect the relationship between the electron and the observer. However, it is not clear how many of these properties might be fundamentally unified and yet differentiate themselves in certain situations only because of a particular relation to an observer. Moreover, what is this electron-observer relation, and how does it give rise to the physics that we are familiar with? Whatever the result of such inquiry, we can be certain that we will have to abandon one or more firmly held concepts related to "particle properties" in favor of something more fundamental.

## 10.3  Influence

Perhaps instead of focusing on what an electron *is* and what properties an electron *possesses*, it would be better to simply focus on what an electron *does* to an observer to define its *relationship* to that observer. Since we know for certain that electrons can influence our equipment and that our equipment can influence electrons, we focus on this simple fact and introduce the concept of *influence*, referring to the resulting theoretical framework as *Influence Theory*.

Inspired in part by Wheeler and Feynman [54, 55], we begin with the simple assumption: *an elementary particle, such as an electron, can influence another particle in a direct and discrete fashion.* One might be tempted to refer to this as a direct particle–particle interaction. However, the term interaction implies something bi-directional, whereas influence here is assumed to be directional—something directed from source to target. As such, each *instance of influence* enables one to define two events: *the act of influencing* and *the act of being influenced.* The act of influencing is associated with the source particle, and the act of being influenced is associated with the target particle.

We imagine the observer to possess an instrument that can monitor an elementary particle and provide information about that particle to the observer. When we say that the observer was influenced by the electron, what we will mean is that the monitored particle was influenced by the electron and that the observer detects the fact that the monitored particle was influenced. Since the aim is to describe the electron, we will take its perspective and refer to the act of influencing associated with the electron as an *outgoing or emitted influence event*. Similarly, the monitored particle may influence the electron. This also is assumed to be detectable, and again, taking the perspective of the electron, we will refer to this act of being influenced associated with the electron as an *incoming or received influence event*.

At this point, the reader is most likely questioning the nature of such influence. We should immediately dispel any notion of propagation and state emphatically that we do not assume that influence propagates through space and time from source to target. Instead, influence simply relates one particle to another in the sense that one particle influenced and the other was influenced. By assuming that only the occurrence of influence is detectable, any properties an instance of influence may possess remain inaccessible. What is remarkable is that we will find that the potentially inaccessible nature of influence does not matter, since what one may erroneously think of as properties of an electron will be shown to emerge as unique consistent descriptions of the influence-based relationship between the electron and the observer. As a result, we will demonstrate that this proposed concept of influence is potentially responsible for the traditional concepts of position, time, motion, energy, momentum, and mass, as well as several well-known quantum effects.

Here we summarize the postulates that provide the foundation for the model.

**Postulate 1** *Elementary particles can influence one another in a pairwise, directed, and discrete fashion, such that given an instance of influence, one particle influences, and the other is influenced.*

This postulate allows us to define the concept of an event.

**Definition 10.1** (*Event*) Every instance of influence results in two *events*, each associated with a different particle: *the act of influencing* and *the act of being influenced*.

By virtue of the fact that these two events can be distinguished, they can be ordered. Consistently choosing the way in which we order the two influence events defines a binary ordering relation $<_i$ that acts on pairs of events, each pair defined by a single influence instance. The subscript $i$ indicates that this ordering is due to an influence instance.

The following postulates together enable one to order the events associated with a single particle. To accomplish this, we assume that the particle has some internal, potentially inaccessible, state that is somehow coupled to the influence instances. We keep this minimal by assuming no details about their relationship. We simply assume that a relationship exists.[1]

**Postulate 2** *A particle has associated with it a potentially inaccessible internal state such that each influence event couples one particle state to one other particle state. It is in this sense that each influence event is bounded by two particle states.*

We also assume that the influence instances are coupled to particle states.

---

[1] If all one can detect is the occurrence of an influence event, how can anything ever be known about the relation between those events and any internal states of the particle?

**Postulate 3** *Each particle state that bounds an influence event couples that influence event to one another influence event associated with the same particle, such that each particle state is bounded by two influence events.*

These two postulates allow one to totally order the influence events associated with a particle, as well as the particle states, with a transitive binary ordering relation, which we shall denote as $<_c$ where the subscript $c$ denotes coupling through internal particle states. This results in one being able to describe a particle in terms of a totally ordered chain of events.[2]

A further consequence of these postulates is the fact that each event along a particle chain is either the result of that particle influencing another or of that particle being influenced. We can take this further and define a new ordering relation $<$ based on considering the two ordering relations $<_i$ and $<_c$ to belong to an equivalence class.

**Definition 10.2** *(Generic Ordering)* Two events $x$ and $y$ are *ordered* with $x$ included by $y$, denoted by $x < y$, if $x <_i y$ or $x <_c y$, or some transitive combination of $<_i$ and $<_c$, such that the arbitrary directions of the binary ordering relations $<_i$ and $<_c$ are selected to avoid cyclic relationships. We can more generally write $x \leq y$ where either $x < y$ or $x = y$.

Together, these postulates result in a model of particle behavior summarized by a set of events, which can be compared using a transitive binary ordering relation defined by the process of influence. This results in an acyclic graph, or a partially-ordered set (poset for short). If one conceives of the ordering as being the foundation of causality, then this is analogous to a causal set [7, 50, 51] where the events are causally ordered, but with a specific connectivity. In this framework, a given particle is described by an ordered sequence, or chain, of events. Each event on one particle chain either covers[3] or is covered by precisely one event on another particle chain. We do not assume that these events take place in any kind of space or time—only that they can be partially ordered.

Given this purposefully simplistic model of a particle, such as an electron, we proceed by developing all aspects of the theory from the bottom up. This work combines the results of several previous efforts that rely on the consistent quantification of systems based on algebraic symmetries [20, 21, 31–37, 53], and we will refer the reader to these works for more details. The idea is to work through each step in sufficient detail to illustrate how quantification of order-theoretic structures enables one to derive laws that reproduce a surprising amount of physics. That is, rather than postulating laws and perceiving them as representing some kind of underlying natural order, we instead postulate the nature of the underlying order and derive the resulting laws.

---

[2]If the particle states were accessible, then we could alternatively describe the particle as a totally ordered chain of particle states.

[3]An event $z$ covers an event $x$ if $x < z$ and there does not exist any $y$ such that $x < y$ and $y < z$.

## 10.4   Coarse-Grained Picture of Influence

### 10.4.1   Intervals: Duration and Directed Distance

It has been posited that an observer monitoring a particle detects events in an ordered sequence, which one can think of as pre-time (ordering without scale). Rather than focusing on the precise poset of events generated by a set of particles influencing one another, we begin by considering a coarse-grained picture of a poset. This is accomplished by defining an order-preserving map that takes a set of successive events along a particle chain to a single element.[4] This will result in a poset of coarse-grained events, each of which consists of multiple fundamental events along with multiple influences. The point of this will be to demonstrate that there exists a unique means (up to scale) by which an embedded observer represented by a chain of events in the poset can quantify a subset of poset and that this is equivalent to the mathematics of space-time.

#### 10.4.1.1   Quantification of a Chain

We begin by introducing the concept of quantification, which will allow us to quantify events and their relationships to one another using numbers. In general, a consistent quantification of the events comprising the observer chain (either fine-grained or coarse-grained) is given by a monotonic valuation, which is a real-valued function $q$ that acts as an order-isomorphism taking each event to a real number such that if the event $y$ includes the event $x$, $x \leq y$, then the real-valued quantities $q(x)$ and $q(y)$ assigned to events $x$ and $y$, respectively, are related by $q(x) \leq q(y)$ (see Fig. 10.1a). The idea is that the quantification captures the ordering of the events experienced by the observer.

#### 10.4.1.2   Incomparability and Inaccessibility

If it is true that all elements of a chain $\mathbf{P}$ are incomparable to an event $x$ in the poset,[5] then we say that the event $x$ is *inaccessible* to the chain $\mathbf{P}$. On the other hand, if there exists at least one element of the chain that is comparable to the event $x$, then we say that $x$ is *accessible* to the chain $\mathbf{P}$. This implies that every observer chain divides the poset into two subsets: a set of events that are accessible to the observer and a set of events that are inaccessible to the observer. Thus the universe of events is naturally divided into the observable universe and the unobservable universe.

---

[4]As an example, given events $p_1 < p_2 < p_3 < \cdots < p_{12}$ along the chain $\mathbf{P}$, the map $\phi$ which gives $\phi(p_1) = \phi(p_2) = \phi(p_3) < \phi(p_4) = \phi(p_5) = \phi(p_6) < \phi(p_7) = \phi(p_8) = \phi(p_9) < \phi(p_{10}) = \phi(p_{11}) = \phi(p_{12})$ is a valid coarse-graining map.

[5]The event $x$ is said to be incomparable to the event $y$ if it is true that $x \not\leq y$ and $y \not\leq x$.

**Fig. 10.1  a** A quantification of a finite chain is performed by assigning a monotonic valuation to the events. In this case, we simply count the events using natural numbers. **b** Any event from the set of accessible events can be quantified with respect to an observer by either one or two numbers by forward projecting and backward projecting onto the observer chain (depending on whether those projections exist)

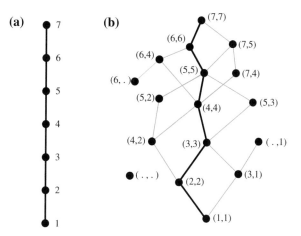

### 10.4.1.3   Chain Projection and the Quantification of Accessible Events

We can extend the concept of quantification of an observer chain to the set of accessible events by means of assigning up to two unique numbers representing the relationship between an accessible event $x$ and the observer chain $\mathbf{P}$.

If the event $x$ is included by an event on the observer chain, that is, if there exists some element $p \in \mathbf{P}$ such that $x \le p$, then we can define a mapping $P$ called the *forward projection* that takes $x$ to the least element of $\mathbf{P}$ that includes it, which is given by $Px = \inf\{p \in P \mid x \le p\}$. This allows us to quantify the event $x$ by assigning to it the quantity assigned to the element $Px$ on the chain. Similarly, if the event $x$ includes an event on the observer chain, such that there exists some element $p \in \mathbf{P}$ such that $p \le x$, then we can define a mapping $\bar{P}$ called the *backward projection* that takes $x$ to the greatest element of $\mathbf{P}$ that it includes, which is given by $\bar{P}x = \sup\{p \in P \mid p \le x\}$. This provides a second possible means by which one can quantify the event $x$ by assigning it the quantity assigned to the element $\bar{P}x$ on the chain. This means that any event accessible to the observer $\mathbf{P}$ can be uniquely quantified by either one or two numbers resulting in a rather strange observer-based coordinate system (Fig. 10.1b).

### 10.4.1.4   Quantification of Intervals

We can extend this concept of consistent quantification to intervals, which are defined by a pair of events $a$ and $b$ (either comparable or incomparable) and denoted $[a, b]$.[6] If we restrict ourselves to the special case where both the forward projection and backward projection of both $a$ and $b$ onto the observer chain $\mathbf{P}$ exist, then the forward and

---

[6]The two events defining the interval are assumed to be collinear to the coordinated pair of observers. This is precisely defined in [36] in terms of projections.

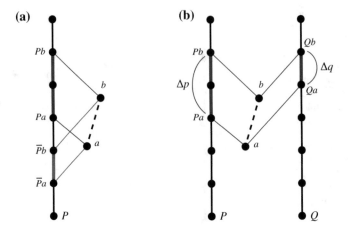

**Fig. 10.2** **a** Intervals, which describe the relationship between two events, can be uniquely quantified by chain projection as well. In this example, the interval can be quantified by four numbers, $Pa, Pb, \bar{P}a, \bar{P}b$, determined by the projections of the events defining the endpoints of the interval onto the observer chain; a pair of numbers, $(Pb - Pa, \bar{P}b - \bar{P}a)$, which reflect the lengths of the projections of the interval onto the observer chain; and a scalar, $(Pb - Pa)(\bar{P}b - \bar{P}a)$, given by the product of those lengths. **b** Two coordinated observers can be used to quantify an interval using forward projections only. This also allows one to define a discrete $1 + 1$-dimensional subspace in the poset. Here the interval is quantified by four numbers, $Pa, Pb, Qa, Qb$; a pair of numbers $(Pb - Pa, Qb - Qa) = (\Delta p, \Delta q)$; and a scalar $(Pb - Pa)(Qb - Qa) = \Delta p \Delta q$, such that the pair and scalar quantifications agree with those obtained using both forward and backward projections onto a single chain as in **a**

backward projections of the pair of events $a$ and $b$ take the interval $[a, b]$ to intervals $[Pa, Pb]$ and $[\bar{P}a, \bar{P}b]$, respectively, on the observer chain (Fig. 10.2a). One can prove [36] that this results in three unique consistent (with respect to rescaling) ways to quantify an interval with respect to an observer:

| | | |
|---|---|---|
| Quadruple | $(Pa, Pb, \bar{P}a, \bar{P}b)$ | (10.1) |
| Pair | $(Pb - Pa, \bar{P}b - \bar{P}a)$ | (10.2) |
| Scalar | $(Pb - Pa)(\bar{P}b - \bar{P}a)$ | (10.3) |

where the pair is comprised of the lengths of the projected intervals on the observer chain, and the scalar is the product of those lengths.

One can reasonably assume that an observer can only obtain information about an electron if it is influenced by that electron. This suggests that one can only quantify observer-accessible events by using forward projection. This gives rise to the concept of a coordinated pair of observers, which is a pair of observers that influence one another in consistent fashion such that the two observers agree on the lengths of intervals on each other's chains. This turns out to be equivalent to defining a $1 + 1$-dimensional inertial frame [36]. Quantification of an interval $[a, b]$ by a pair of coordinated observers **P** and **Q** using forward projections is illustrated in Fig. 10.2b. Again, the interval can be consistently quantified in three ways[7]:

| | | |
|---|---|---|
| Quadruple | $(Pa, Pb, Qa, Qb)$ | (10.4) |
| Interval Pair | $(\Delta p, \Delta q)$ | (10.5) |
| Interval Scalar | $\Delta p \, \Delta q$ | (10.6) |

where $\Delta p = Pb - Pa$ and $\Delta q = Qb - Qa$ are the projected lengths of the intervals on the observer chains.

One can transform to a more convenient set of coordinates by considering symmetric and antisymmetric combinations of projected interval lengths:

$$\Delta t = \frac{\Delta p + \Delta q}{2} \tag{10.7}$$

$$\Delta x = \frac{\Delta p - \Delta q}{2}. \tag{10.8}$$

The quantity $\Delta t$, which is referred to as *duration*, quantifies intervals that lie along the observer chains (an ordered relationship), and the quantity $\Delta x$, which can be referred to as a *directed distance*, quantifies the relationships between chains (an unorderable relationship).[8] These quantities, $\Delta t$ and $\Delta x$, are related to the interval scalar (10.6) by

$$\Delta s^2 = \Delta p \, \Delta q \tag{10.9}$$

$$= \left(\frac{\Delta p + \Delta q}{2}\right)^2 - \left(\frac{\Delta p - \Delta q}{2}\right)^2 \tag{10.10}$$

$$\equiv \Delta t^2 - \Delta x^2, \tag{10.11}$$

which is the Minkowski metric in $1 + 1$ dimensions. Thus the mathematics of space and time appears to emerge as the unique means by which embedded observers can quantitatively describe the events accessible to them. We refer to this observer-based description of the poset as the *space-time picture*.

### 10.4.2   Motion and Velocity

In the previous section, we demonstrated that concepts such as duration and directed distance reflect the relationship between the events and the observers. Consequently, and perhaps not unexpectedly, these quantities are expected to change when one transforms from one pair of observers to another.

---

[7]Please see [36] for technical details.

[8]Directed distance differs from distance by at most a sign, which indicates the orientation of the interval with respect to the observers **P** and **Q**.

One can consider transforming from one pair of coordinated observers **PQ** to a second pair of coordinated observers **P'Q'** in the more general case where intervals of length $\kappa$ along the chains **P** and **Q** project to intervals of length $m$ on **P'** and intervals of length $n$ on **Q'**. For the observers to consistently quantify intervals with the interval scalar, we have that

$$\kappa^2 = mn, \tag{10.12}$$

which implies that the interval pair transforms as [36]

$$(\Delta p', \Delta q') = \left( \sqrt{\frac{m}{n}} \Delta p, \sqrt{\frac{n}{m}} \Delta q \right). \tag{10.13}$$

This implies that the quantity $\Delta t$, quantifying duration along chains, and the quantity $\Delta x$, quantifying directed distance between chains, will mix when transforming to the primed coordinate system. It is a matter of straightforward algebra to show that the transformation

$$\Delta t' = \frac{\sqrt{\frac{m}{n}} + \sqrt{\frac{n}{m}}}{2} \Delta t + \frac{\sqrt{\frac{m}{n}} - \sqrt{\frac{n}{m}}}{2} \Delta x \tag{10.14}$$

$$\Delta x' = \frac{\sqrt{\frac{m}{n}} - \sqrt{\frac{n}{m}}}{2} \Delta t + \frac{\sqrt{\frac{m}{n}} + \sqrt{\frac{n}{m}}}{2} \Delta x \tag{10.15}$$

results in the Lorentz transformation

$$\Delta t' = \frac{1}{\sqrt{1 - \beta^2}} \Delta t + \frac{\beta}{\sqrt{1 - \beta^2}} \Delta x \tag{10.16}$$

$$\Delta x' = \frac{\beta}{\sqrt{1 - \beta^2}} \Delta t + \frac{1}{\sqrt{1 - \beta^2}} \Delta x \tag{10.17}$$

consistent with special relativity where

$$\beta = \frac{m - n}{m + n}. \tag{10.18}$$

The quantity $\beta$ is immediately recognized as the velocity of the unprimed frame **PQ** with respect to the primed frame **P'Q'**. This is because $m$ and $n$ are the projected lengths, $\Delta p'$ and $\Delta q'$, of an interval of length $\kappa$ on chains **P** and **Q** onto **P'** and **Q'**. The quantity $\frac{m-n}{2}$ is the directed distance of that interval in the primed frame, and $\frac{m+n}{2}$ is the projected duration in the primed frame, so that in general

$$\beta = \frac{\Delta p' - \Delta q'}{\Delta p' + \Delta q'} \equiv \frac{\Delta x'}{\Delta t'}, \tag{10.19}$$

as expected.

One of the strange results of special relativity is the fact that durations and distances are not concrete fixed physical quantities, but rather they are observer-dependent and can change. In the context of Influence Theory, this is not strange at all, since the quantities of duration, directed distance, and velocity merely reflect the relationship between the observer and the observed, and changing observers would, in general, be expected to result in a change in these quantities. As a consequence, these results would suggest that space and time are not physical things with properties. Space and time are instead uniquely consistent descriptions of the relationship between the observed and the observer.

### 10.4.3   Rates: Energy, Momentum and Mass

We have been looking at events along the observer chain in terms of intervals, which led to relevant variables that reflect the concepts of duration, directed distance, and velocity. With any ordered sequence of elements, there is a dual perspective where one describes the sequence in terms of rates. Intervals and rates are Fourier transforms of one another and as such, the interval scalar, duration, and directed distance each have Fourier duals, which we show are related to mass, energy, and momentum, respectively.

We define the rates at which a particle influences a pair of coordinated observers in terms of a selected number $N$ of outgoing influence events emitted by the particle chain divided by the duration as measured by an observer.

$$r_P = \frac{N}{\Delta p} \qquad r_Q = \frac{N}{\Delta q}. \tag{10.20}$$

As such, these rates are, again, observer-based. Combining them in a symmetric and antisymmetric fashion [32, 34] results in the quantities that we will refer to as *energy*

$$E = \frac{r_P + r_Q}{2}, \tag{10.21}$$

*momentum*

$$p = \frac{r_Q - r_P}{2}, \tag{10.22}$$

and *mass*

$$m = \sqrt{r_P r_Q} \tag{10.23}$$

such that the familiar mass-energy-momentum relationship holds:

$$m^2 = E^2 - p^2. \tag{10.24}$$

If we assume that the particle has no preference for influencing $\mathbf{P}$ or $\mathbf{Q}$, then we can write

$$\langle \Delta p \rangle = \frac{N}{2}k \qquad \langle \Delta q \rangle = \frac{N}{2}\frac{1}{k}, \qquad (10.25)$$

where the factor $k = \sqrt{\frac{m}{n}}$ from (10.13) reflects the choice of scale $k = 1$ in the rest frame and naturally ensures Lorentz invariance. We can then write the mass squared as

$$m^2 = r_P r_Q = \frac{N^2}{\langle \Delta p \rangle \langle \Delta q \rangle} = 4 \qquad (10.26)$$

giving us a mass[9] of

$$m = 2. \qquad (10.27)$$

Keep in mind that at this point, this is a single particle theory. We have not yet considered a model where particles can influence at different rates.

It is straightforward to verify that the velocity, as defined in (10.19), can be written in terms of energy (10.21) and momentum (10.22) as

$$\beta = \frac{p}{E}. \qquad (10.28)$$

Furthermore, since $m^2 = r_P r_Q$ is an invariant, it is also straightforward to verify that these rate-based quantities transform properly under boosts [32, 34].

While the mass-energy-momentum relation (10.24) appears here in a new foundational context, such a conception of mass, energy and momentum should not be surprising as it is closely related to the concept of the *internal electron clock rate* hypothesized by de Broglie [12, 25]. In his 1924 thesis, de Broglie considered Planck's Law, which already considers energy to be a frequency,

$$E = \hbar \omega, \qquad (10.29)$$

and Einstein's Law, which relates energy to mass,

$$E = mc^2, \qquad (10.30)$$

and reasoned that mass was related to a frequency

$$m = \frac{\hbar \omega}{c^2}. \qquad (10.31)$$

There is no mystery here. Duration and energy, and directed distance (position) and momentum, are *not complementary properties* of an electron, which cannot be measured accurately simultaneously. They are instead *complementary descriptions of*

---

[9]This observation was made by James L. Walsh.

*the relationship* between the electron and the observer. Duration and directed distance are obtained by considering intervals, whereas energy and momentum are obtained by considering rates, which are necessarily long-term averages. This aspect of quantum complementarity, along with its reliance on the Fourier transform, emerges naturally as a relationship between the means by which one describes sequences of events.

## 10.5   Fine-Grained Picture of Influence

Up until this point, we have been considering a coarse-grained picture of influence, which has resulted in well-defined concepts of duration, directed distance, energy, momentum, and mass. We now return to the fine-grained picture, where every event on the particle chain represents either an act of influence directed at another particle or the act of being influenced by another particle. The aim is to apply the concepts developed in the previous sections to better understand how a particle, such as an electron, should behave at the most fundamental level from the perspective of Influence Theory.

### 10.5.1   Zitterbewegung

As is usual in physics, we gain insight by focusing on an ideal situation. Traditionally, a free particle is a particle that is free from the influence of outside forces. Here we define a *free particle* as a particle that influences others but is itself not influenced. As before, we focus on the case of $1 + 1$ dimensions as defined by two coordinated observers where the particle is assumed to be collinear to these observers (see [36] for technical details).

Figure 10.3a illustrates a free particle $\Pi$ that influences the chains $\mathbf{P}$ and $\mathbf{Q}$. Unlike in the coarse-grained case each event on the particle chain is covered by only one event on either the chain $\mathbf{P}$ or the chain $\mathbf{Q}$. Forward projecting that event to the chain that it influences is trivial. However, forward projecting to the other chain relies on transitivity through some successive event (if it exists). This allows one to project intervals defined by successive events on $\Pi$ onto the observers $\mathbf{P}$ and $\mathbf{Q}$ as illustrated in Fig. 10.3.

One gains significant insight by considering the velocities assigned to the intervals along the particle chain defined by successive events. Note that any such interval forward projects to one chain resulting in a projected interval of non-zero length, while forward projecting to the other chain resulting in a degenerate interval of zero length (see Fig. 10.3b). In the case where the lesser event on the particle chain influences $\mathbf{P}$, we then have that $\Delta p > 0$ and $\Delta q = 0$. Since the velocity is given by

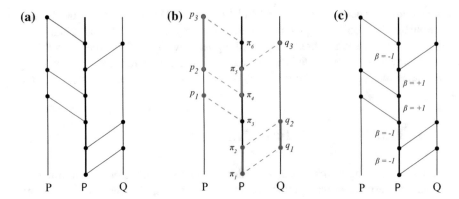

**Fig. 10.3** **a** A free particle $\Pi$ that influences the observer chains **P** and **Q**. Note that the physical distance between events on the sheet of paper is meaningless—it is necessary for the particle chain to be drawn as a straight line. **b** An illustration of the projection of intervals defined by successive events on the chain $\Pi$. The interval $[\pi_1, \pi_2]$ projects onto a degenerate interval $[p_1, p_1]$ of length $\Delta p = 0$ on the chain **P** and projects onto the interval $[q_1, q_2]$ of length $\Delta q > 0$ on the chain **Q**. Similarly, the interval $[\pi_4, \pi_5]$ projects onto the interval $[p_2, p_3]$, which has length $\Delta p > 0$, and the degenerate interval $[q_3, q_3]$ with length $\Delta q = 0$. **c** The lengths of the projections of these intervals along the particle chain onto the observer chains allow one to define an associated velocity. Here we see that the velocities $\pm 1$ are assigned to each successive interval

$$\beta = \frac{\Delta x}{\Delta t} \tag{10.32}$$

$$= \frac{\Delta p - \Delta q}{\Delta p + \Delta q} \tag{10.33}$$

we have that $\beta = +1$. Similarly, in the case where the lesser event on the particle chain influences **Q**, we have that $\Delta p = 0$ and $\Delta q > 0$ so that $\beta = -1$. Thus when the particle influences **P**, the observers describe it as moving to the right at the ultimate speed. Similarly, when the particle influences **Q**, the observers describe it as moving to the left at the ultimate speed.

A free particle influencing to the right and left with non-zero rates, $r_P$ and $r_Q$, has a mass $m = \sqrt{r_P r_Q}$. In this case, every observer describes the particle as zig-zagging back-and-forth at the speed of light, with probabilities of moving left and right given by

$$Prob(R) \equiv Prob(\beta = +1) = \frac{\Delta p}{\Delta p + \Delta q} \tag{10.34}$$

$$= \frac{r_Q}{r_P + r_Q} \tag{10.35}$$

and

$$\text{Prob}(L) \equiv \text{Prob}(\beta = -1) = \frac{\Delta q}{\Delta p + \Delta q} \tag{10.36}$$

$$= \frac{r_P}{r_P + r_Q}. \tag{10.37}$$

One can find the average velocity, $\langle \beta \rangle$, by

$$\langle \beta \rangle = \text{Prob}(R) - \text{Prob}(L), \tag{10.38}$$

or equivalently one can write the probabilities in terms of the average velocity as

$$\text{Prob}(R) = \frac{1 + \langle \beta \rangle}{2} \tag{10.39}$$

and

$$\text{Prob}(L) = \frac{1 - \langle \beta \rangle}{2}. \tag{10.40}$$

Note that the rates $r_P$ and $r_Q$ transform inversely under boosts, so that the mass is invariant. However, this changes the probabilities with which the particle is observed by a given observer to move left or right, which results in a change of average velocity $\langle \beta \rangle$, as well as energy and momentum, as expected when one keeps in mind that these are all observer-based quantities reflecting a description of the relationship between the particle and observer. Despite the fact that it has been demonstrated that this behavior is consistent with special relativity on average [33], it is curious that not even special relativity can describe what the universe looks like to these particles at the finest of scales, since relativity does not describe what happens when a particle moves at the speed of light.

It is interesting to consider the limiting case of a massless particle, which requires that one of the influence rates be zero. That is, either $r_P = 0$ or $r_Q = 0$ so that $m = \sqrt{r_P r_Q} = 0$ and $E = |p|$.[10] In this case, the particle does not *zitter*. Instead, it influences only to the left or only to the right, which is described by observers as a massless particle traveling in one direction at the constant speed of light. However, to obtain a rate of zero, the particle chain itself must project to a degenerate interval on one of the observer chains.

In relativistic quantum mechanics, this behavior of zig-zagging back-and-forth at the speed of light is predicted by the Dirac equation. This predicted phenomenon was originally noted by Breit [8] and Schrödinger [47], the latter of whom coined the term *Zitterbewegung*, or 'shuddering motion', which we will shorten to *zitter*. The phenomenon of *zitter* was later emphasized by Huang [27], discussed by Feshbach and

---

[10]It is important to note that the case where both rates are zero would result in not only zero mass but also zero energy and momentum. Such a particle would not influence anything and would therefore be unobserved.

Villars [14], described in a handful of texts [5, 6, 15, 40], and has been championed by Hestenes [23–26], Barut [2–4] and others [22, 45, 46, 49] who have conceived of *zitter* as a spiral motion and envisioned a connection to spin. In the present context of $1 + 1$ dimensions, *zitter* manifests as a discrete zig-zagging motion, which, as we will discuss below, is central to the Feynman checkerboard model of the electron [15]. While *zitter* has not yet been observed directly in electrons, there is not only indirect evidence that this is a real physical effect for electrons [10, 19] but also evidence that this is a real effect for fermions in general [17, 39, 44, 56].

The phenomenon of *zitter* should be of some concern since it challenges the concept of rest by implying that at the most fundamental level every particle is in a constant state of motion at the speed of light. That is, all particles, massive and massless, can *only* go the speed of light! All other speeds—including rest—are observed only on average. As a result, *zitter* goes as far as challenging the concept of a rest frame, which is central to the theory of general relativity. This suggests that a theory such as general relativity can only hold on average in a coarse-grained picture and thus is likely to be inconsistent with quantum mechanics.

### 10.5.2   Influence Sequences: Further Quantum Effects

The proposal that a particle's position is defined, at least in part, by its influence on coordinated observers is a novel perspective, which may at first glance seem to run contrary to the usual idea that observers must make measurements, which affect the particle's behavior, to learn about a particle. However, in this section we will show that the proposed model gives rise to several more quantum effects, including information isolation and the fact that measurements disturb particles.

The concept of position was derived as a consistent description of intervals between events in a coarse-grained setting. We have seen in the previous section on *Zitterbewegung*, that the description of intervals defined by influence events at the fundamental level leads to unexpected results that are consistent with some very poorly understood aspects of relativistic quantum mechanics. Here we explore further consequences of the model.

#### 10.5.2.1   Compton Wavelength

We begin by assuming that one knows the initial state of a free particle and consider the changes in the position (10.8) and the time (10.7) used to describe the particle after it influences one of the observers.[11] When the observer **P** is influenced, both the position and the time describing the particle change by $\frac{\Delta p}{2}$. Similarly, influencing **Q** changes the position by $-\frac{\Delta q}{2}$, and the time by $\frac{\Delta q}{2}$. In the rest frame, for any

---

[11]By 'position' and 'time', we mean the directed distance and duration with respect to a defined origin.

influence event, either $\Delta p = 1$ or $\Delta q = 1$. Thus the time coordinate describing a particle advances in a discrete fashion by $\frac{1}{2}$, and its position can change by $\pm\frac{1}{2}$. This means that the discrete nature of the act of influence imposes a fundamental unit of duration and distance beyond which one cannot measure. One might postulate that this fundamental distance is related to the reduced Compton wavelength

$$\lambda = \frac{\hbar}{mc}, \tag{10.41}$$

which in natural units reduces to

$$\lambda = \frac{1}{m}. \tag{10.42}$$

Given the fact that $m = 2$ (10.27), we see that this is indeed the case. The reduced Compton wavelength is simply the smallest definable distance

$$\lambda = \frac{1}{2}. \tag{10.43}$$

### 10.5.2.2   Information Isolation and the Unorderability of Influence Sequences

We now consider two coordinated observers, **P** and **Q**, monitoring a free particle, represented by the chain of events $\Pi$. The particle is assumed to be collinear to **PQ**, so that we are considering a $1 + 1$ dimensional space induced by the two observers. The observers record influence events and aim to make inferences about the sequence in which the particle has influenced the observers, which we refer to as an *influence sequence*. However, despite the fact that both **P** and **Q** have carefully recorded and quantified the influence events on their own chains, they are fundamentally unable to deduce the order in which the influence events occur on the particle chain $\Pi$. This can be demonstrated by considering a simple example where more than one particle chain can give rise to the same influence events on the observer chains.

We consider a free particle that evolves from an initial state, represented by $X$, to a final state by influencing the observers **P** and **Q** at the events labeled $p_1$, $p_2$, and $q_1$ in Fig. 10.4. In this case, there are three possible orderings of the events on the free particle chain $\Pi$ that result in the same situation for the two observers. In the first case, we have that the particle $\Pi$ influences **P** at $\pi_1$, **P** at $\pi_2$, and **Q** at $\pi_3$, which we write compactly as the influence sequence $[X, P, P, Q]$. The other two possible sequences, $[X, P, Q, P]$ and $[X, Q, P, P]$, are illustrated in Fig. 10.4. In general, one can show that the number of influence sequences to be considered is given by

$$S = \binom{N_p + N_q}{N_p} = \binom{N_p + N_q}{N_q} = \frac{(N_p + N_q)!}{N_p!N_q!}, \tag{10.44}$$

where $N_p$ and $N_q$ represent the number of times that the particle influences **P** and **Q**.

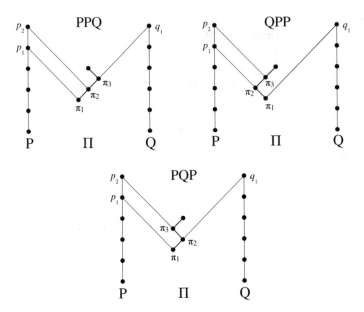

**Fig. 10.4** Given just the influence events $p_1$, $p_2$, and $q_1$ associated with the observers **P** and **Q**, there are three possible ways in which they could have resulted from the three acts of influence, $\pi_1$, $\pi_2$, and $\pi_3$ associated with the particle $\Pi$. As a result, when making inferences about the behavior of the particle $\Pi$ as it evolves from its initial state $x_i$ (associated with $\pi_1$) to the final state (unlabeled), the observers must consider all possible orderings of influence instances

Moreover, since each influence sequence of the particle corresponds to a different description of the particle by the observers, each distinct influence sequence corresponds to a distinct space-time path. This is illustrated in Fig. 10.4 where the chain $\Pi$ is drawn specifically to depict the resulting space-time path in each case. As a result, the fact that the observers cannot determine the precise influence sequence of a free particle means that they cannot ascribe to that free particle a particular path through space-time. Any inferences that the observers attempt to make about the free particle's behavior must take into account the set of possible influence sequences or, equivalently, the set of possible space-time paths. This set of possible influence sequences constitute a set of *interfering alternatives* [15].

The fact that the observers are fundamentally unable to determine the precise influence sequence experienced by the free particle is an example of a key characteristic of quantum systems known as *information isolation* [48]. Here, the information about the order in which the particle influences observers **P** and **Q** is inherently isolated from the observers. There is no way for the observers to order these events. Thus it is fundamentally impossible to describe the particle behavior precisely. In this model, it is not that the free particle takes all possible paths through space-time. Instead, because of information isolation, the observers must consider all possible paths to make optimal inferences about the particle. The only question that remains is how does one consistently take this information (or lack of information) into account to make such optimal inferences.

### 10.5.2.3 Measurement

The fact that the influence sequence of a free particle is informationally isolated from the observers results in requiring that the observers consider all possible influence sequences when making inferences about the particle. The situation changes if the observers themselves influence the particle. Figure 10.5 illustrates a particle that is influenced by the observer **Q** at event $\pi_4$ while it influences the observers. The act of influencing the particle enables the observers to order some of the influence events along the particle chain $\Pi$. For example, in this case it is known that event $q_2 \leq p_3$. This constrains the possible set of influence sequences associated with the particle chain and at least partially destroys the information isolation by providing some information about the particle behavior. However, in doing so, the observer necessarily influenced the particle. We refer to this process as *measurement*, as it results in providing information about the particle to the observers while affecting the particle behavior as one would expect in a quantum mechanical system.

### 10.5.2.4 Quantifying the Free Particle State

In this section we aim to describe how one can go about describing the free particle state, or influence sequence, for the purposes of making inferences. Given that the free particle is fundamentally described by an influence sequence, we must focus

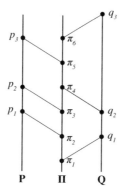

**Fig. 10.5** This figure illustrates the process of measurement where by influencing the particle at the event $\pi_4$, the observer **Q**, at least temporarily, disrupts information isolation and obtains information about the particle behavior. In this case, the information comes in the form of a constraint, where it is learned that $q_2 \leq p_3$. This limits the number of influence sequences that must be considered to make inferences about the particle $\Pi$

on describing transitions from one influence event to another. This results in four possible transitions:

$$P \rightarrow P \qquad Q \rightarrow P \qquad (10.45)$$
$$P \rightarrow Q \qquad Q \rightarrow Q. \qquad (10.46)$$

We quantify each of these transitions with a complex number. This is a big step and is justified, in part, by symmetries described in previous works [20, 21], which have been shown to be applicable to this problem [32]. Here we present an alternative perspective that is perhaps more intuitive.

We aim to quantify the behavior of the particle for the purposes of making inferences. That is, we will assign a number or numbers to particular particle behaviors with the aim of consistently computing probabilities. We have been considering the behavior of a free particle in terms of intervals defined by the transition from one influence event to another, which can be described by observers as a particle taking discrete steps in a space-time. Alternatively, we could consider the behavior of the free particle in terms of rates, which can be described in terms of energy and momentum. The two perspectives are Fourier duals to one another. For this reason, the quantities that we use to describe the particle in these dual perspectives must also be Fourier duals to one another. The simplest quantities that are assured to work in general are complex numbers. However, there is a constraint that regardless of the perspective chosen, we must make consistent inferences. That is, the probabilities computed using complex numbers describing influence sequence transitions must be equal to the probabilities computed from the Fourier transform of those complex numbers, which describe the influence sequence rates. Parseval's theorem [1] states that the only scalar derived from complex numbers that is invariant under the Fourier transform is the squared magnitude. This is the Born Rule, simply conceived, as a consistency requirement arising from the fact that we must make consistent inferences when describing influence sequences in terms both of intervals and rates.

The fact that we are quantifying transitions from one influence event to another in a finite set of influence sequences implies that we must account for the fact that we will not, in general, possess complete information about any chosen 'initial state'. For this reason, we need to consider both the possibility that the particle has previously influenced **P** and the possibility that it has previously influenced **Q**. These two situations, which are analogous to arriving at the initial state from the left and right, respectively, are the two helicity states. To quantify them, we require two complex numbers:

$$\Phi = \begin{pmatrix} \phi_{\mathbf{P}} \\ \phi_{\mathbf{Q}} \end{pmatrix}. \qquad (10.47)$$

The complex numbers $\phi_{\mathbf{P}}$ and $\phi_{\mathbf{Q}}$ are the quantum amplitudes for the particle to influence **P** and **Q**, respectively. The probability that the particle influences **P** is then given by

$$\phi_{\mathbf{P}}^{*}\phi_{\mathbf{P}} \equiv \mathrm{Prob}(R)$$

$$= \frac{1 + \langle \beta \rangle}{2}$$

$$= \frac{E + p}{2E}. \tag{10.48}$$

Similarly, the probability that the particle influences $\mathbf{Q}$ is given by

$$\phi_{\mathbf{Q}}^{*}\phi_{\mathbf{Q}} = \frac{E - p}{2E}. \tag{10.49}$$

This allows us to write these numbers in a more familiar form

$$\phi_{\mathbf{P}} = \sqrt{\frac{E + p}{2E}} \, e^{i\theta_P} \quad \phi_{\mathbf{Q}} = \sqrt{\frac{E - p}{2E}} \, e^{i\theta_Q} \tag{10.50}$$

where $\theta_P$ and $\theta_Q$ are phase angles.

The result is that to make inferences about the influence patterns of a free particle, the observers must describe the particle with two complex numbers that take the form of a Pauli spinor.[12]

In addition, it is important to note that since the state of a particle is defined in terms of the way in which the particle influences the observers, the fact that a particular space-time coordinate can be arrived at in only two ways, by influencing $\mathbf{P}$ or by influencing $\mathbf{Q}$, implies that at most two particles can occupy the same space-time coordinates and that these two particles must have opposite helicity states. For example, consider the space-time position associated with the event $\pi_3$ in Fig. 10.4 in both the QPP and PQP cases. In the QPP case, the particle arrives at the space-time position associated with $\pi_3$ by influencing $\mathbf{P}$ at the event $p_1$. Whereas, in the PQP case, the particle arrives at the same space-time position by influencing $\mathbf{Q}$ at the event $q_1$. Since that space-time position can only be reached by influencing $p_1$ or $q_1$, and since each event on the observer chain is defined by the influence of one and only one particle, this implies that no more than two particles can arrive at the same space-time position and that they must arrive with different helicity states. Thus this model results in a $1 + 1$-dimensional version of the *Pauli Exclusion Principle*.

### 10.5.3 The Feynman Checkerboard Model and the Dirac Equation

In 1965, Feynman and Hibbs [15] introduced the discrete one-dimensional checkerboard model of the electron where massive particles are conceived of as moving at

---

[12]Our initial studies of influenced particles indicate that one needs four complex numbers and that they appear to take the form of a Dirac spinor with the positive energy components representing the amplitudes for the particle to influence and the negative energy components representing the amplitudes for the particle to be influenced.

the speed of light, but zig-zagging back-and-forth resulting in a subluminal average velocity. This is typically illustrated in the standard space-time diagram as a path that zig-zags upwards along 45° angle light cones.

In the checkerboard model, each path segment is assigned a quantum amplitude so that the total amplitude of a given path (path amplitude) is found by taking the product of the amplitudes assigned to each segment comprising the path. The amplitude assigned to the transition from some initial state at time $t_i$ and position $x_i$ to some final state at time $t_f$ and position $x_f$ is found by the discrete path integral summing the path amplitudes over all paths connecting $(t_i, x_i)$ to $(t_f, x_f)$. It is posed as Problem 2–6 in Feynman and Hibbs [15, p. 35] to derive the Dirac equation by assigning an amplitude of $i\varepsilon$ to path reversals and an amplitude of $\varepsilon$ to straightaways.

Given the way in which space-time coordinates are assigned to influence events in this current theory, one can see that this model is isomorphic to the Feynman checkerboard (see Fig. 10.4 where the particle chain is illustrated as following a space time path). Summing the amplitudes of space-time paths is equivalent to summing the amplitudes for all possible influence sequences. We have previously shown that the amplitude assignments proposed by Feynman and Hibbs can be *derived* using symmetries and consistency with probability theory [32] in terms of transition matrices, or propagators, that evolve the spinor as the particle influences **P** or **Q**:

$$P = \frac{1}{\sqrt{2}} \begin{pmatrix} 1 & i \\ 0 & 0 \end{pmatrix} \quad Q = \frac{1}{\sqrt{2}} \begin{pmatrix} 0 & 0 \\ i & 1 \end{pmatrix}, \tag{10.51}$$

so that

$$P\Phi = \frac{1}{\sqrt{2}} \begin{pmatrix} 1 & i \\ 0 & 0 \end{pmatrix} \begin{pmatrix} \phi_P \\ \phi_Q \end{pmatrix} \tag{10.52}$$

$$= \frac{1}{\sqrt{2}} \begin{pmatrix} \phi_P + i\phi_Q \\ 0 \end{pmatrix} \tag{10.53}$$

and

$$Q\Phi = \frac{1}{\sqrt{2}} \begin{pmatrix} 0 & 0 \\ i & 1 \end{pmatrix} \begin{pmatrix} \phi_P \\ \phi_Q \end{pmatrix} \tag{10.54}$$

$$= \frac{1}{\sqrt{2}} \begin{pmatrix} 0 \\ i\phi_P + \phi_Q. \end{pmatrix}. \tag{10.55}$$

One can then readily derive the Dirac equation for the free particle as

$$\partial_P \phi_P = i\phi_Q \tag{10.56}$$

$$\partial_Q \phi_Q = i\phi_P, \tag{10.57}$$

which can be rewritten as

$$(\partial_t + \partial_x)\phi_P = i\,m\,\phi_Q \tag{10.58}$$
$$(\partial_t - \partial_x)\phi_Q = i\,m\,\phi_P, \tag{10.59}$$

where the mass $m = 2$ as was found earlier (10.27).

Observables can then be represented as matrix operators. For example, the average velocity $\langle\beta\rangle$ of a particle can be found by defining the operator

$$\hat{\beta} = \begin{pmatrix} 1 & 0 \\ 0 & -1 \end{pmatrix} \tag{10.60}$$

so that

$$\begin{aligned}
\langle\beta\rangle &= \Phi^\dagger\hat{\beta}\Phi \\
&= (\phi_P{}^* \;\; \phi_Q{}^*)\begin{pmatrix} 1 & 0 \\ 0 & -1 \end{pmatrix}\begin{pmatrix} \phi_P \\ \phi_Q \end{pmatrix} \\
&= (\phi_P{}^* \;\; \phi_Q{}^*)\begin{pmatrix} \phi_P \\ -\phi_Q \end{pmatrix} \\
&= \phi_P^*\phi_P - \phi_Q^*\phi_Q.
\end{aligned} \tag{10.61}$$

This can also be written as $Prob(R) - Prob(L)$ as derived in (10.38), or alternatively, by using (10.48) and (10.49), this can be written as $p/E$.

Here we see that the Feynman checkerboard model of the electron, which has been a curiosity of sorts attracting attention now-and-again throughout the years [13, 16, 18, 28, 30, 38, 41, 42], is isomorphic to the simple model of an elementary particle that influences others in a direct and discrete fashion.

## 10.6 Discussion

The field of physics has been slowly constructed over the last four hundred years by identifying principles and relevant variables that aid in the optimal prediction of physical phenomena. Since the discovery of the electron, physicists have struggled with the fact that several aspects of the mental models we use to conceive of an electron appear to be logically inconsistent despite the fact that the optimal predictive theory, known as quantum electrodynamics (QED), employs accepted relevant variables along with adopted principles to make the most accurate predictions of any physical theory in history. As a result, many foundational theorists work to develop sets of logically consistent principles by which quantum theory can be reconstructed. While it is perhaps accepted that a successful theory will most likely need to revise or discard one or more commonly held beliefs, many foundational approaches attempt to retain as many familiar concepts and technical assumptions as possible so as to ensure success. While this is a wise approach in some respects, it is not assured that it will result in what one might hope for in terms of a truly foundational theory. This is because by assuming the relevance of a specific variable or adopting a specific principle or technical description, one is prevented from learning more about them. For this reason, we have adopted a different foundational approach: build physics from the bottom up.

Rather than postulating laws and perceiving them as representing some kind of underlying natural order, we instead postulate the nature of the underlying order and derive the resulting laws. This is accomplished by considering physics to represent a framework by which observers consistently quantify and make consistent optimal inferences about natural phenomena. To do this successfully, it is important to look for clues. From our previous efforts [20, 21, 31–37, 53], and those of others [9, 11, 29, 43], we have learned that symmetries inherent to a system constrain any attempt to consistently quantify that system. That is, it is possible to begin with an underlying order and derive laws that are consistent with that order via the process of *consistent quantification*. This is the reason why mathematics is so successful at describing physics [35]. This is not idle philosophy. We apply this critical observation to a simple model of an electron and use it to construct (reconstruct) a consistent physical theory from the bottom up.

We observe that the majority of the variables relevant to an electron which are often conceived of as representing properties of an electron, represent instead the relationship between the electron and an observer. Based on this observation, we introduce the concept of influence as a simple means by which the electron-observer relationship is mediated. It is postulated that an elementary particle, such as an electron, can influence another particle in a direct and discrete fashion. Each instance of influence enables one to define two events, each associated with a different particle: the act of influencing and the act of being influenced. Two additional postulates result in influence events being considered as a partially ordered set. This allows us to describe a given particle as a totally ordered chain of events.

Consistent quantification of intervals defined by pairs of influence events results in the mathematics of space-time. This implies that space and time need not be physical. Instead, space and time are the uniquely consistent constructs by which one can describe events from the perspective of an embedded observer. By considering events along a particle chain in terms of rates, we recover the concepts of mass, energy, and momentum, and note that these are necessarily Fourier duals of the invariant interval scalar, duration, and directed distance (position). As a result the concept of complementarity emerges—not as complementarity among properties of particles, but rather as complementary descriptions of particles.

The fact that intervals and rates are Fourier duals suggests that the mathematical formalism of quantum mechanics might be derived by considering probabilities to be computed from quantities describing influence sequences. The fact that one's inferences should be invariant with respect to one's description of a system in terms of intervals or rates, suggests that such systems should be quantified by complex numbers and that the Born Rule is simply an example of Parseval's theorem applied to influence sequences.

In addition to complementarity, several other concepts central to quantum mechanics emerge naturally from the model: information isolation, the Compton wavelength, and the Pauli exclusion principle. Optimal inferences about the behavior of a free particle result in the Dirac equation with Pauli spinors used to quantify the particle behavior in $1 + 1$-dimensions. The proposed model exhibits *Zitterbewegung*, which is a poorly understood relativistic quantum effect predicted by the Dirac equation and intimately related to mass, spin, and velocity. It is presently thought that the Higgs field gives rise to mass. However, this does not explain the intimate relationship between mass and spin, which has been investigated by Hestenes and others. From the perspective of the proposed theory, *Zitterbewegung*, mass, and (at least) helicity (a $1 + 1$-dimensional analog of spin) arise from the fact that particles influence one another. This leads one to wonder if the Higgs field simply represents this network of influence instances.

Despite the fact that a surprising amount of physics can be derived from this simple model of one particle influencing another, it would be naive to assume that the ideas presented here comprise anything resembling the final word on the matter. It is hoped that this work has demonstrated that a simple and understandable picture of particles, such as the electron, as well as a broad and coherent foundational theory of physics are indeed feasible and should be actively sought after.

**Acknowledgements** I would like to thank Newshaw Bahreyni, Seth Chaiken, Ariel Caticha, Keith Earle, David Hestenes, Oleg Lunin, John Skilling, and James Lyons Walsh for numerous insightful discussions. I also want to specifically thank James Lyons Walsh for his careful proofreading of this manuscript and his invaluable comments.

# References

1. Arfken, G.: Mathematical Methods for Physicists. Academic Press, Orlando, FL (1985)
2. Barut, A.O., Bracken, A.J.: Magnetic-moment operator of the relativistic electron. Phys. Rev. D **24**, 3333–3334 (1981)
3. Barut, A.O., Bracken, A.J.: Zitterbewegung and the internal geometry of the electron. Phys. Rev. D **23**, 2454–2463 (1981)
4. Barut, A.O., Zanghi, N.: Classical model of the Dirac electron. Phys. Rev. Lett. **52**, 2009–2012 (1984)
5. Baym, G.A.: Lectures on Quantum Mechanics. Addison-Wesley (1969)
6. Bjorken, J.D. Drell, S.D.: Relativistic Quantum Mechanics. McGraw-Hill (1964)
7. Bombelli, L., Lee, J.H., Meyer, D., Sorkin, R.: Space-time as a causal set. Phys. Rev. Lett. **59**, 521–524 (1987)
8. Breit, G.: An interpretation of Dirac's theory of the electron. Proc. Nat. Acad. Sci. **14**(7), 553 (1928)
9. Caticha, A.: Consistency, amplitudes, and probabilities in quantum theory. Phys. Rev. A **57**(3), 1572–1582 (1998)
10. Catillon, P., Cue, N., Gaillard, M.J., Genre, R., Gouanère, M., Kirsch, R.G., Poizat, J.C., Remillieux, J., Roussel, L., Spighel, M.: A search for the de Broglie particle internal clock by means of electron channeling. Found. Phys. **38**(7), 659–664 (2008)

11. Cox, R.T.: Probability, frequency, and reasonable expectation. Am. J. Phys. **14**, 1–13 (1946)
12. De Broglie, L.: Recherches sur la théorie des quanta. Ph.D. thesis, Migration-université en cours d'affectation (1924)
13. Earle, K.A.: A master equation approach to the '3 + 1' Dirac equation. (arXiv:1102.1200 [math-ph]) (2011)
14. Feshbach, H., Villars, F.: Elementary relativistic wave mechanics of spin 0 and spin 1/2 particles. Rev. Mod. Phys. **30**(1), 24 (1958)
15. Feynman, R.P., Hibbs, A.R.: Quantum mechanics and path integrals. McGraw-Hill, New York (1965)
16. Gaveau, B., Schulman, L.S.: A projector path integral for the Dirac equation and the spin derivation of space. Ann. Phys. **284**(1), 1–9 (2000)
17. Gerritsma, R., Kirchmair, G., Zähringer, F., Solano, E., Blatt, R., Roos, C.F.: Quantum simulation of the Dirac equation. Nature **463**(7277), 68–71 (2010)
18. Gersch, H.A.: Feynman's relativistic chessboard as an Ising model. Int. J. Theor. Phys. **20**(7), 491–501 (1981)
19. Gouanère, M., Spighel, M., Cue, N., Gaillard, M.J., Genre, R., Kirsch, R., Poizat, J.C., Remillieux, J., Catillon, P., Roussel, L.: Experimental observations compatible with the particle internal clock. Annales de la Fondation Louis de Broglie **30**(1), 109–14 (2005)
20. Goyal, P., Knuth, K.H.: Quantum theory and probability theory: their relationship and origin in symmetry. Symmetry **3**(2), 171–206 (2011)
21. Goyal, P., Knuth, K.H., Skilling, J.: Origin of complex quantum amplitudes and Feynman's rules. Phys. Rev. A **81**, 022109, (arXiv:0907.0909 [quant-ph]) (2010)
22. Gull, S., Lasenby, A., Doran, C.: Electron paths, tunnelling, and diffraction in the spacetime algebra. Found. Phys. **23**(10), 1329–1356 (1993)
23. Hestenes, D.: The Zitterbewegung interpretation of quantum mechanics. Found. Phys. **20**(10), 1213–1232 (1990)
24. Hestenes, D.: Zitterbewegung modeling. Found. Phys. **23**(3), 365–387 (1993)
25. Hestenes, D.: Electron time, mass and zitter. "The Nature of Time" FQXi 2008 Essay Contest (2008)
26. Hestenes, D.: Zitterbewegung in quantum mechanics. Found. Phys. **40**(1), 1–54 (2010)
27. Huang, K.: On the Zitterbewegung of the Dirac electron. Am. J. Phys. **20**, 479 (1952)
28. Jacobson, T.: Feynman's checkerboard and other games. In: Sanchez, N. (ed.) Non-Linear Equations in Classical and Quantum Field Theory, pp. 386–395. Springer, Berlin Heidelberg (1985)
29. Jaynes, E.T.: The evolution of Carnot's principle. In: Erickson, G.J., Smith, C.R. (eds.) Maximum Entropy and Bayesian Methods in Science and Engineering, vol. 1, pp. 267–281. Springer (1988)
30. Kauffman, L.H., Noyes, H.P.: Discrete physics and the Dirac equation. Phys. Lett. A **218**(3), 139–146 (1996)
31. Knuth, K.H.: Deriving laws from ordering relations. In: Zhai, Y., Erickson, G.J. (eds.) Bayesian Inference and Maximum Entropy Methods in Science and Engineering, pp. 204–235. Jackson Hole WY, USA, August 2003, AIP Conf. Proc. 707, AIP, New York, (arXiv:physics/0403031v1 [physics.data-an]) (2004)
32. Knuth, K.H.: Information-based physics: an observer-centric foundation. Contemporary Phys. **55**(1), 12–32, (arXiv:1310.1667 [quant-ph]) (2014)
33. Knuth, K.H.: The problem of motion: the statistical mechanics of Zitterbewegung. In: Mohammad-Djafari, A., Barbaresco, F. (eds.) Bayesian Inference and Maximum Entropy Methods in Science and Engineering. Amboise, FRANCE, AIP, New York (2014)
34. Knuth, K.H.: Information-based physics and the influence network. In: Aguirre, A., Foster, B., Merali, Z. (eds.) Bit or Bit from It? On Physics and Information, pp. 65–78. Springer Frontiers Collection, Springer, Heidelberg, FQXi 2013 Essay Contest (Third Prize), (arXiv:1308.3337 [quant-ph]) (2015)
35. Knuth, K.H.: The deeper roles of mathematics in physical laws. In press, FQXi 2015 Essay Contest (Third Prize), (arXiv:1504.06686 [math.HO]) (2016)

36. Knuth, K.H., Bahreyni, N.: A potential foundation for emergent space-time. J. Math. Phys. **55**, 112501, (arXiv:1209.0881 [math-ph]) (2014)
37. Knuth, K.H., Skilling, J.: Foundations of inference. Axioms **1**(1), 38–73 (2012)
38. Kull, A.: Quantum mechanical motion of relativistic particle in non-continuous spacetime. Phys. Lett. A **303**(2), 147–153 (2002)
39. LeBlanc, L.J., Beeler, M.C., Jimenez-Garcia, K., Perry, A.R., Sugawa, S., Williams, R.A., Spielman, I.B.: Direct observation of zitterbewegung in a Bose–Einstein condensate. New J. Phys. **15**(7), 073011 (2013)
40. Merzbacher, E.: Quantum Mechanics (1998)
41. Ord, G.N., Gualtieri, J.A.: The Feynman propagator from a single path. Phys. Rev. Lett. **89**(25), 250403–250406 (2002)
42. Ord, G.N., McKeon, D.G.C.: On the Dirac equation in 3+1 dimensions. Ann. Phys. **222**(2), 244–253 (1993)
43. Pfanzagl, J.: Theory of Measurement. Wiley (1968)
44. Qu, C., Hamner, C., Gong, M., Zhang, C., Engels, P.: Observation of Zitterbewegung in a spin-orbit-coupled Bose-Einstein condensate. Phys. Rev. A **88**(2), 021604 (2013)
45. Rodrigues, Jr. W.A., Vaz, Jr. J., Recami, E., Salesi, G.: About Zitterbewegung and Electron Structure (1998)
46. Salesi, G., Recami, E.: Field theory of the spinning electron and internal motions. Phys. Rev. A **190**(2), 137–143 (1994)
47. Schrödinger, E.: Über die kräftefreie bewegung in der relativistischen quantenmechanik. Akademie der wissenschaften in kommission bei W. de Gruyter u, Company (1930)
48. Schumacher, B., Westmoreland, M.: Quantum Processes, Systems, and information. Cambridge University Press (2010)
49. Sidharth, B.G.: Revisiting zitterbewegung. Int. J. Theor. Phys. **48**(2), 497–506 (2009)
50. Sorkin, R.D.: Causal sets: discrete gravity. In: Gomberoff, A., Marolf, D. (eds.) Lectures on Quantum Gravity, pp. 305–327. Springer US, (arXiv:gr-qc/0309009) (2005)
51. Sorkin, R.D.: Geometry from order: causal sets. In: Einstein Online, vol. 2, 1007, http://www.einstein-online.info/spotlights/causal_sets (2006)
52. Thomson, J.J.: Cathode rays. The Electrician **39**, 104–109 (1897)
53. Walsh, J.L., Knuth, K.H.: Information-based physics, influence, and forces. In: Mohammad-Djafari, A., Barbaresco, F. (eds.) Bayesian Inference and Maximum Entropy Methods in Science and Engineering. Amboise, France, AIP Conf. Proc., AIP, New York (2014)
54. Wheeler, J.A., Feynman, R.P.: Interaction with the absorber as the mechanism of radiation. Rev. Mod. Phys. **17**(2–3), 157–181 (1945)
55. Wheeler, J.A., Feynman, R.P.: Classical electrodynamics in terms of direct interparticle action. Rev. Mod. Phys. **21**(3), 425–433 (1949)
56. Zawadzki, W., Rusin, T.M.: Zitterbewegung (trembling motion) of electrons in semiconductors: a review. J. Phys. Condensed Matt. **23**(14), 143201 (2011)

# Index

© Springer International Publishing Switzerland 2017
I.T. Durham and D. Rickles (eds.), *Information and Interaction*,
The Frontiers Collection, DOI 10.1007/978-3-319-43760-6

Printed in the United States
By Bookmasters